SpringerBriefs in Materials

The SpringerBriefs Series in Materials presents highly relevant, concise monographs on a wide range of topics covering fundamental advances and new applications in the field. Areas of interest include topical information on innovative, structural and functional materials and composites as well as fundamental principles, physical properties, materials theory and design. SpringerBriefs present succinct summaries of cutting-edge research and practical applications across a wide spectrum of fields. Featuring compact volumes of 50 to 125 pages, the series covers a range of content from professional to academic. Typical topics might include:

- A timely report of state-of-the-art analytical techniques
- A bridge between new research results, as published in journal articles, and a contextual literature review
- A snapshot of a hot or emerging topic
- An in-depth case study or clinical example
- A presentation of core concepts that students must understand in order to make independent contributions

Briefs are characterized by fast, global electronic dissemination, standard publishing contracts, standardized manuscript preparation and formatting guidelines, and expedited production schedules.

More information about this series at http://www.springer.com/series/10111

Waseem A. Wani · Mohammad Shahid
Afzal Hussain · Mohamed Fahad AlAjmi

Fluorescent Organic Nanoparticles

New Generation Materials with Diverse
Analytical and Biomedical Applications

 Springer

Waseem A. Wani
Department of Chemistry
Government Degree College Tral
Tral, Jammu and Kashmir, India

Mohammad Shahid
Centre for Textile Conservation
    and Technical Art History
University of Glasgow
Glasgow, UK

Afzal Hussain
Department of Pharmacognosy,
    College of Pharmacy
King Saud University
Riyadh, Saudi Arabia

Mohamed Fahad AlAjmi
Department of Pharmacognosy,
    College of Pharmacy
King Saud University
Riyadh, Saudi Arabia

ISSN 2192-1091   ISSN 2192-1105   (electronic)
SpringerBriefs in Materials
ISBN 978-981-13-2654-7   ISBN 978-981-13-2655-4   (eBook)
https://doi.org/10.1007/978-981-13-2655-4

Library of Congress Control Number: 2018955163

This Springer imprint is published by the registered company Springer Nature Singapore Pte Ltd.
The registered company address is: 152 Beach Road, #21-01/04 Gateway East, Singapore 189721, Singapore

Preface

Fluorescent organic nanoparticles (FONs) are a very interesting class of materials with diverse analytical and biomedical applications. Several FONs have been reported as attractive agents for cell imaging applications. Over the last few decades, several bioprobes, e.g. organic dyes, fluorescent proteins and fluorescent inorganic/organic nanoparticles, have been reported for biomedical applications. FONs are thought as more promising agents for biomedical applications owing to their possibility of diverse designs and biodegradability properties. Drug delivery systems have brought about a great revolution in the pharmaceutical field. The use of drug delivery systems can help in avoiding some of the inherent drawbacks of the commonly used drugs, such as low solubility in physiological systems, leaching, lower activity, unwanted interactions with different biological macromolecules other than the target ones, toxicity, and decomposition. Photoresponsive nanoparticles have been the preferred choice for drug delivery applications owing to their ability to control the release of pharmacologically important drugs via externally regulated stimulation of light. In the light of these facts, FONs have also been extensively studied as sensors in drug delivery systems and for other applications like photodynamic therapy and apoptosis inducers of cancer cells.

Nowadays, design and development of highly sensitive and selective fluorescent probes for sensing biologically important analytes in aqueous or cellular environments is an active area of research. Several types of fluorescent materials, such as small organic dyes conjugated with polymers, organic nanoparticles, inorganic quantum dots, metallic nanoclusters and upconversion nanoparticles, have been used for sensing different types of analytes including cationic and anionic ones.

The recent research carried out on the design and development of aggregation-induced emission (AIE) nanoparticles is a promising stimulation towards the development of highly fluorescent nanoparticles as the natural aggregation is kept busy in increasing the fluorescence of the nanoparticles. AIE is therefore truly a novel finding of this subject. The encapsulation of an emitter into different matrices affects their aggregation, molecular packing and distribution in the nanoparticles. Thus, care is needed while selecting the polymeric matrix and the emitter-to-matrix ratio, which may help in the tuning of nanoparticle size, brightness and stability. Several reports have been published with polymer-encapsulated emitters having sizes ranging from few to several hundred nanometres. Such nanoparticles have versatile surface functional groups that have been tailored to provide space for different imaging needs, for example imaging of cellular organelles, targeted *in vitro* and *in vivo* imaging of tumours, tracing of cancer cells, and imaging of blood vessels and specific chemical and biomolecular species.

The recent exploration of the cell imaging, chemosensing and drug delivery applications of FONs has brought about a great revolution. This is very important and interesting from the materials point of view. Nevertheless, there has been a good deal of work in the development of highly fluorescent nanoparticles for cell imaging, sensing and drug delivery applications; several existing FONs are not very emissive in the aggregated states. Thus, FONs with strong near-infrared (NIR) absorption and active non-radiative emission are the materials of choice at the moment. The non-radiative pathway is generally associated with heat production, and thus, such FONs might have great potential for use in photothermal therapy, wherein highly specific, non-toxic and non-invasive treatments of cancers may be carried out. It is of great interest that several FONs have also been tried as agents for photodynamic therapy of cancers; however, this field of research is at an early stage as the principles of design of such FONs and their action mechanisms are not fully established. However, such nanoparticulate systems have successfully confirmed their promise for the deep in vivo tumour imaging and photodynamic therapeutic uses with minimum or no side effects on normal cells and tissues. The future research on FONs needs to be focused on the design and development of smart stimuli-responsive sensing, imaging and drug delivery systems. Besides, the future work must focus on developing FONs and AIE fluorogens with far red/near-infrared (FR/NIR) emitters displaying narrow band emission for specific analytical and biomedical applications.

In this book, attempts have been made to address the advances made in the development of FONs as materials of choice for the design and fabrication of sensors, bioimaging agents and drug delivery systems. Basically, four important methods, namely self-assembly, polymerization, emulsification and nanoprecipitation/reprecipitation, have been used for the preparation of FONs with diverse applications in analytical and biomedical sciences. Out of these techniques, nanoprecipitation is the simplest and the most widely used technique. This technique enables the transformation of soluble organic molecules into nanoparticles in the aqueous media and later ensures their fast screening for various analytical and biomedical applications. We have also tried our best to throw light on the outlooks of the research and development in FONs as smart materials with various possible applications. In the context of fluorescence-based sensing, drug delivery and cell imaging, the properties of FONs that are of major importance include their stability, brightness, toxicity and biodegradability. Overall, FONs represent a very interesting field of research with promise for varied applications in analytical and biomedical sciences.

Tral, India   Dr. Waseem A. Wani
Glasgow, UK   Dr. Mohammad Shahid
Riyadh, Saudi Arabia   Dr. Afzal Hussain
Riyadh, Saudi Arabia   Dr. Mohamed Fahad AlAjmi

Contents

About the Authors

Dr. Waseem A. Wani is currently working as an Assistant Professor at the Department of Chemistry, Govt. Degree College Tral, Kashmir, India. He completed his PhD at Jamia Millia Islamia, New Delhi, India in 2014 with a thesis on "Syntheses, characterization and anti-cancer profiles of glutamic acid derivatives and their metal ion complexes." Dr. Wani's research areas include Organic and Inorganic Syntheses, Synthesis of Nanomaterials, Chromatography, Anticancer Metallodrugs, Nanofunctionalization of Anticancer Metallodrugs, and Natural Product Chemistry. He has authored or co-authored more than 40 research publications (original research papers, reviews, book chapters, books, editorials and conference papers).

Dr. Mohammad Shahid received master's degree in Chemistry from Shibli National College, Azamgarh (India) in 2006 and Ph.D. in Organic Chemistry from the Jamia Millia Islamia, New Delhi (India) in 2014 under the supervision of Dr. Faqeer Mohammad. He is currently working as Marie Skłodowska-Curie Fellow at the University of Glasgow, UK. He has published more than 30 papers in international journals and conference proceedings, and is an active reviewer for many prominent journals.

Dr. Afzal Hussain is an Assistant Professor at the Department of Pharmacognosy, College of Pharmacy, King Saud University, Riyadh, Saudi Arabia. He completed his PhD in Chemistry at Jamia Millia Islamia, New Delhi, India in 2010. Having been admitted as a Member to the Royal Society of Chemistry (MRSC) in 2018, he has eight years of experience in teaching and research. His areas of research include separation sciences, bioanalysis, drug analysis, cosmeceutical analysis, nutraceutical analysis, and synthesis of novel nanomaterials as drug delivery carriers. Dr. Hussain has published more than 40 papers in international journals and conference proceedings, and is an active reviewer for many prominent journals.

Dr. Mohamed Fahad AlAjmi is Head of the Skills Development Unit at the Department of Pharmacognosy, College of Pharmacy, King Saud University, Riyadh, Saudi Arabia. He completed his PhD in Pharmacy in 2007. Dr. AlAjmi has published more than 60 papers in international journals and is a member of many professional societies, including the American Society for Mass Spectrometry and American Society of Pharmacognosy. He has received two prestigious awards from King Saud University: the Excellence Award in Skills Development (2013) and the Creative Learning and Teaching Award (2016).

Chapter 1
Introduction—Fluorescence in Organic Nanoparticles

The noticeable advantages of the readiness of biologically compatible imaging agents, pilotable instruments and high progressive resolution with good sensitivity have made the applications of fluorescence imaging techniques very interesting [1–3]. For obtaining preferred signal output and high signal-to-noise ratio, several discrete molecules and colloidal nanoparticles have been used as fluorescent probes. Fluorescent organic nanoparticles (FONs) are a very important class of materials; composed of small organic molecules with emissive properties comparable to those of semi-conjugated polymer dots [4].

Generally, FONs contain a high number of photo-active self-assembled units (commonly greater than 10^5 per FON). Besides, there are no stabilizing surfactants or doping matrices within the structures of FONs, which helps them to avoid the undesirable interventions of non-active matter with living organisms. These nanoparticles show intense brightness upon one- or two-photon excitation, which helps in achieving reduced excitation fluence and tissue auto-fluorescence [5, 6]. It is due to the peculiar features of FONs (good photostability and facile surface functionalization) that have made them attractive agents in biomedical research [7]. Several classes of organic and inorganic nanomaterials and some organic-inorganic hybrids have been explored as fluorescent probes for applications in in vitro and in vivo sensing and imaging [8–14]. Out of the several classes of materials developed so far, organic nanoparticles containing organic emitters as fluorescent cores have been thought as promising probes for in vivo applications due to their high photoluminescence, biodegradability and flexible synthetic approaches [15, 16]. The broad range of applications such as in opto-electronic nanodevices, bio- and chemical sensing, drug delivery and monitoring systems, diagnostics, immunofluorescent labelling and in vitro/in vivo imaging of FONs has made their design and development as one of the rapidly growing fields of organic nanochemistry [17–22]. Easy cellular uptake [23], size-dependence of fluorescent properties [15] and longer fluorescence lifetimes [24] are some of the most exciting features of FONs.

From the literature updates, it can be seen that several FONs, e.g. fluorescent conjugated polymers [25–33], self-assembled fluorescent nanoparticles [33–36], poly-

© The Author(s), under exclusive license to Springer Nature Singapore Pte Ltd. 2018
W. A. Wani et al., *Fluorescent Organic Nanoparticles*, SpringerBriefs in Materials,
https://doi.org/10.1007/978-981-13-2655-4_1

dopamine nanoparticles and aggregation-induced-emission (AIE) or aggregation-induced emission enhancement (AIEE) nanoparticles have been reported recently [37–39]. Importantly, AIE or AIEE FONs have shown amazing anti-aggregation-caused quenching (anti-ACQ) in comparison to the traditional organic dyes. Besides, several AIE or AIEE moieties including siloles [40], cyano-substituted diarylethene [18, 41], tetraphenylethene [42–44], triphenylethene [45, 46] and distyrylanthracene derivatives [47–49] have been explored for bio-imaging and chemosensing applications. FONs are more promising for biomedical applications than fluorescent inorganic nanoparticles owing to their flexible design and biodegradability. One of the most applicable approaches for the fabrication of FONs is the controlled self-assembly of monodisperse π-conjugated oligomers or chromophores [34, 50, 51]. Nowadays, FONs are being explored as chemical sensors, biosensors, photosensitizers and cell imaging agents in aqueous media [52–57]. Due to the appreciable stability of FONs in aqueous medium, these materials provide convenient routes for chemical and biological investigations in aqueous solutions [58, 59].

An exhaustive literature search through SciFinder indicated around 206 research papers on "fluorescent organic nanoparticles" with applications in cell imaging, sensing and drug delivery, among others. An observation of the available literature on FONs indicated that interest in the research on organic nanoparticles with fluorescence properties has increased steadily during the last two decades owing to their applications in different analytical techniques and healthcare systems. The increasing research interest in this field can be seen from Fig. 1.1, which indicates that the number of research papers published annually is continuously increasing from 2006. Also to our observations, no review/book is available in the literature on FONs. There is only one partially relevant review by Li and Liu [60], wherein the authors have discussed advances in fluorescence and photo-acoustic imaging applications of organic nanoparticles encapsulated in polymers. Till date, there is no reference material on the advances in FONs with applications in sensing, cell imaging and drug delivery. Therefore, this book has been written to address the advances in the development of FONs as sensing and cell imaging agents and drug delivery agents. This book is expected to become a useful and informative reference material for researchers and academics involved in the research and development of novel materials based on FONs.

1.1 Fluorescence in Organic Nanoparticles

The presence of weak interactions like Van der Waals forces and hydrogen bonding between organic molecules differentiates between their properties and those of inorganic nanostructures [61, 62]. The fascination of researchers towards nanoparticles is because they show interesting optical properties such as size-dependent absorption, emission spectra [63–66], piezochromic luminescence [67] and AIE [68–72]. Excitation-dependent fluorescence (EDF) is another interesting property of organic nanoparticles [73–75]. EDF has been reported in some inorganic [76–79] and organic

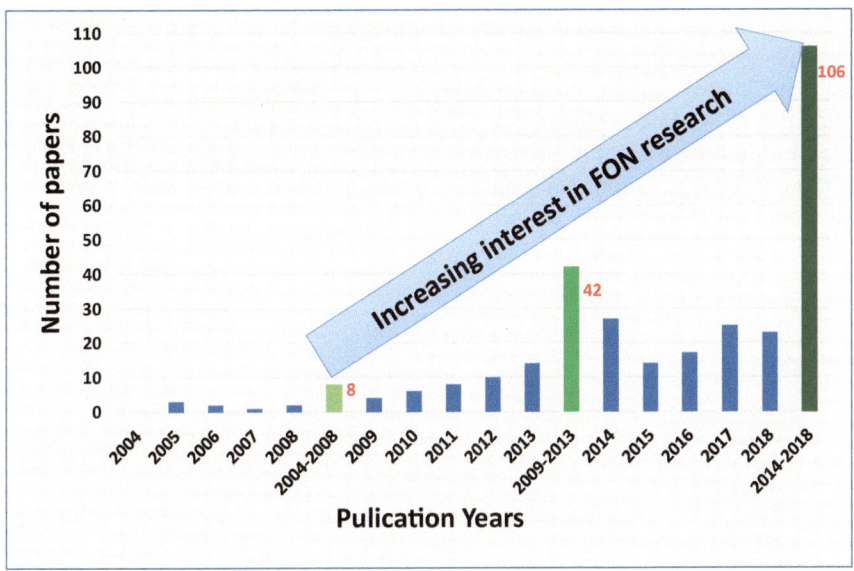

Fig. 1.1 A histogram representation of the increasing attention towards research in FONs

nanostructures [80] with applications in opto-electronics, biosensing and chemical sensing devices. Nanoparticle EDF is an unusual phenomenon as it violates Kasha's rule [81]. Several mechanisms such as incomplete solvation [76], size distribution of nanostructures, [73, 78, 82] and contamination of ions [83] have been put forward to explain nanoparticle EDF. The research on FONs was stimulated by Kasai et al. in 1999 [63, 84]. They found that perylene and phthalocyanine nanoparticles exhibited different and size-dependent fluorescence properties with respect to their bulk samples. Recently, Yao et al. reported similar results with pyrazoline nanoparticles [15]. It also has been known that there is a decrease in fluorescence and photochemical stability of nanoparticles formed by aggregation of fluorescent organic molecules. The decrease in fluorescence and photochemical stability has been primarily attributed to the self-quenching aspect of the individual molecules within the nanoparticle [85–87]. However, it was recently observed that aggregation-induced emission enhancement (AIEE) may be a result of a decrease in fluorescence quenching at high concentrations of fluorophore [88, 89]. Generally, AIEE is known to occur when fluorescent materials emit intensely in their aggregated or solid state with very weak emission in the solution phase. AIEE was reported in molecules with rotor structures, and thus, it is suggested that the attachment of such structures to fluorescent molecules might be a strategy to convert them into AIEE molecules [90–93].

The impact of the effect of aggregation in organic nanoparticles may be either similar to or quite different than what we generally observe in bulk organic materials. The spectral features of organic nanoparticles are generally interpreted in terms of understanding of the effect of molecular aggregation on electronic absorption and emission spectral properties. The relative orientation of the aggregating molecules and hence their resultant transition dipoles have significant effects on the spectral properties of organic nanoparticles. As shown by the excitons coupling mode, a side-by-side arrangement of the transition dipoles (H-aggregate) causes blue shift in absorption with diminished emission intensity. On the other hand, a head-to-tail alignment (J-aggregate) causes a red shift of absorption along with enhanced emission intensity [94].

References

1. Leblond, F., Davis, S.C., Valdes, P.A., Pogue, B.W.: Pre-clinical whole-body fluorescence imaging: review of instruments, methods and applications. J. Photochem. Photobiol., B **98**(1), 77–94 (2010)
2. Kondepati, V.R., Heise, H.M., Backhaus, J.: Recent applications of near-infrared spectroscopy in cancer diagnosis and therapy. Anal. Bioanal. Chem. **390**(1), 125–139 (2008)
3. Kircher, M.F., Gambhir, S.S., Grimm, J.: Noninvasive cell-tracking methods. Nat. Rev. Clin. Oncol. **8**(11), 677–688 (2011)
4. Wu, C., Bull, B., Szymanski, C., Christensen, K., McNeill, J.: Multicolor conjugated polymer dots for biological fluorescence imaging. ACS Nano **2**(11), 2415–2423 (2008)
5. Ishow, E., Brosseau, A., Clavier, G., Nakatani, K., Tauc, P., Fiorini-Debuisschert, C., Neveu, S., Sandre, O., Léaustic, A.: Multicolor emission of small molecule-based amorphous thin films and nanoparticles with a single excitation wavelength. Chem. Mater. **20**(21), 6597–6599 (2008)
6. Parthasarathy, V., Fery-Forgues, S., Campioli, E., Recher, G., Terenziani, F., Blanchard-Desce, M.: Dipolar versus octupolar triphenylamine-based fluorescent organic nanoparticles as brilliant one-and two-photon emitters for (bio) imaging. Small **7**(22), 3219–3229 (2011)
7. De, M., Ghosh, P.S., Rotello, V.M.: Applications of nanoparticles in biology. Adv. Mater. **20**(22), 4225–4241 (2008)
8. Chen, M., Yin, M.: Design and development of fluorescent nanostructures for bioimaging. Prog. Polym. Sci. **39**(2), 365–395 (2014)
9. Smith, A.M., Duan, H., Mohs, A.M., Nie, S.: Bioconjugated quantum dots for in vivo molecular and cellular imaging. Adv. Drug Deliv. Rev. **60**(11), 1226–1240 (2008)
10. Wilcoxon, J.P., Abrams, B.L.: Synthesis, structure and properties of metal nanoclusters. Chem. Soc. Rev. **35**(11), 1162–1194 (2006)
11. Wang, F., Liu, X.: Recent advances in the chemistry of lanthanide-doped upconversion nanocrystals. Chem. Soc. Rev. **38**(4), 976–989 (2009)
12. Baker, S.N., Baker, G.A.: Luminescent carbon nanodots: emergent nanolights. Angew. Chem. **49**(38), 6726–6744 (2010)
13. Wang, L., Zhao, W., Tan, W.: Bioconjugated silica nanoparticles: development and applications. Nano Res. **1**(2), 99–115 (2008)
14. Feng, G., Liang, J., Liu, B.: Hyperbranched conjugated polyelectrolytes for biological sensing and imaging. Macromol. Rapid Commun. **34**(9), 705–715 (2013)
15. Fu, H.-B., Yao, J.-N.: Size effects on the optical properties of organic nanoparticles. J. Am. Chem. Soc. **123**(7), 1434–1439 (2001)
16. An, B.K., Kwon, S.K., Park, S.Y.: Photopatterned arrays of fluorescent organic nanoparticles. Angew. Chem. **46**(12), 1978–1982 (2007)
17. Martin, C.R., Kohli, P.: The emerging field of nanotube biotechnology. Nat. Rev. Drug Discovery **2**(1), 29–37 (2003)

18. An, B.K., Kwon, S.K., Jung, S.D., Park, S.Y.: Enhanced emission and its switching in fluorescent organic nanoparticles. J. Am. Chem. Soc. **124**(48), 14410–14415 (2002)
19. Li, H., Xu, J., Yan, H.: Ratiometric fluorescent determination of cysteine based on organic nanoparticles of naphthalene–thiourea–thiadiazole-linked molecule. Sens. Actuators B: Chem **139**(2), 483–487 (2009)
20. Jana, A., Devi, K.S., Maiti, T.K., Singh, N.D.: Perylene-3-ylmethanol: fluorescent organic nanoparticles as a single-component photoresponsive nanocarrier with real-time monitoring of anticancer drug release. J. Am. Chem. Soc. **134**(18), 7656–7659 (2012)
21. Chandrasekhar, N., Chandrasekar, R.: Reversibly shape-shifting organic optical waveguides: formation of organic nanorings, nanotubes, and nanosheets. Angew. Chem. Int. Ed. **51**(15), 3556–3561 (2012)
22. Jiang, S., Gnanasammandhan, M.K., Zhang, Y.: Optical imaging-guided cancer therapy with fluorescent nanoparticles. J. Roy. Soc., Interface/Roy. Soc. **7**(42), 3–18 (2010)
23. dos Santos, T., Varela, J., Lynch, I., Salvati, A., Dawson, K.A.: Quantitative assessment of the comparative nanoparticle-uptake efficiency of a range of cell lines. Small **7**(23), 3341–3349 (2011)
24. Wang, L., Dong, L., Bian, G., Xia, T., Chen, H., Wang, L., Cao, Q., Li, L.: Application of organic nanoparticles as fluorescence probe in the determination of nucleic acids. Anal. Lett. **37**(9), 1811–1822 (2004)
25. Feng, L., Zhu, C., Yuan, H., Liu, L., Lv, F., Wang, S.: Conjugated polymer nanoparticles: preparation, properties, functionalization and biological applications. Chem. Soc. Rev. **42**(16), 6620–6633 (2013)
26. Zhu, C., Yang, Q., Lv, F., Liu, L., Wang, S.: Conjugated polymer-coated bacteria for multimodal intracellular and extracellular anticancer activity. Adv. Mater. **25**(8), 1203–1208 (2013)
27. Nie, C., Zhu, C., Feng, L., Lv, F., Liu, L., Wang, S.: Synthesis of a new conjugated polymer for DNA alkylation and gene regulation. ACS Appl. Mater. Interfaces **5**(11), 4549–4554 (2013)
28. Wang, B., Zhu, C., Liu, L., Lv, F., Yang, Q., Wang, S.: Synthesis of a new conjugated polymer for cell membrane imaging by using an intracellular targeting strategy. Polym. Chem. **4**(20), 5212–5215 (2013)
29. Lu, X., Jiang, R., Yang, M., Fan, Q., Hu, W., Zhang, L., Yang, Z., Deng, W., Shen, Q., Huang, Y.: Monodispersed grafted conjugated polyelectrolyte-stabilized magnetic nanoparticles as multifunctional platform for cellular imaging and drug delivery. J. Mater. Chem. B **2**(4), 376–386 (2014)
30. Zhang, G., Lu, X., Wang, Y., Huang, Y., Fan, Q., Huang, W.: Water-soluble fluorescent nanoparticles without distinct aggregation of conjugated polymer chains. Polym. Int. **60**(1), 45–50 (2011)
31. Zhang, P., Chen, J., Huang, F., Zeng, Z., Hu, J., Yi, P., Zeng, F., Wu, S.: One-pot fabrication of polymer nanoparticle-based chemosensors for Cu 2+ detection in aqueous media. Polym. Chem. **4**(7), 2325–2332 (2013)
32. Sugihara, S., Armes, S.P., Blanazs, A., Lewis, A.L.: Non-spherical morphologies from cross-linked biomimetic diblock copolymers using RAFT aqueous dispersion polymerization. Soft Matter **7**(22), 10787–10793 (2011)
33. Yin, M., Feng, C., Shen, J., Yu, Y., Xu, Z., Yang, W., Knoll, W., Mullen, K.: Dual-responsive interaction to detect DNA on template-based fluorescent nanotubes. Small **7**(12), 1629–1634 (2011)
34. Cui, K., Lu, X., Cui, W., Wu, J., Chen, X., Lu, Q.: Fluorescent nanoparticles assembled from a poly(ionic liquid) for selective sensing of copper ions. Chem. Commun. **47**(3), 920–922 (2011)
35. Luo, J., Lei, T., Wang, L., Ma, Y., Cao, Y., Wang, J., Pei, J.: Highly fluorescent rigid supramolecular polymeric nanowires constructed through multiple hydrogen bonds. J. Am. Chem. Soc. **131**(6), 2076–2077 (2009)
36. Shi, C., Guo, Z., Yan, Y., Zhu, S., Xie, Y., Zhao, Y.S., Zhu, W., Tian, H.: Self-assembly solid-state enhanced red emission of quinolinemalononitrile: optical waveguides and stimuli response. ACS Appl. Mater. Interfaces **5**(1), 192–198 (2013)

37. Liu, J., Ding, D., Geng, J., Liu, B.: PEGylated conjugated polyelectrolytes containing 2, 1, 3-benzoxadiazole units for targeted cell imaging. Polym. Chem. **3**(6), 1567–1575 (2012)

38. Zhang, X., Zhang, X., Wang, S., Liu, M., Zhang, Y., Tao, L., Wei, Y.: Facile incorporation of aggregation-induced emission materials into mesoporous silica nanoparticles for intracellular imaging and cancer therapy. ACS Appl. Mater. Interfaces **5**(6), 1943–1947 (2013)

39. Leung, C.W., Hong, Y., Chen, S., Zhao, E., Lam, J.W., Tang, B.Z.: A photostable AIE luminogen for specific mitochondrial imaging and tracking. J. Am. Chem. Soc. **135**(1), 62–65 (2013)

40. Luo, J., Xie, Z., Lam, J.W., Cheng, L., Chen, H., Qiu, C., Kwok, H.S., Zhan, X., Liu, Y., Zhu, D., Tang, B.Z.: Aggregation-induced emission of 1-methyl-1,2,3,4,5-pentaphenylsilole. Chem. Commun. **18**, 1740–1741 (2001)

41. Zhang, X., Ma, Z., Liu, M., Zhang, X., Jia, X., Wei, Y.: A new organic far-red mechanofluorochromic compound derived from cyano-substituted diarylethene. Tetrahedron **69**(49), 10552–10557 (2013)

42. Xue, W., Zhang, G., Zhang, D., Zhu, D.: A new label-free continuous fluorometric assay for trypsin and inhibitor screening with tetraphenylethene compounds. Org. Lett. **12**(10), 2274–2277 (2010)

43. Zhang, X., Chi, Z., Li, H., Xu, B., Li, X., Liu, S., Zhang, Y., Xu, J.: Synthesis and properties of novel aggregation-induced emission compounds with combined tetraphenylethylene and dicarbazolyl triphenylethylene moieties. J. Mater. Chem. **21**(6), 1788–1796 (2011)

44. Zhang, X., Chi, Z., Zhang, Y., Liu, S., Xu, J.: Recent advances in mechanochromic luminescent metal complexes. J. Mater. Chem. C **1**(21), 3376–3390 (2013)

45. Zhang, X., Yang, Z., Chi, Z., Chen, M., Xu, B., Wang, C., Liu, S., Zhang, Y., Xu, J.: A multi-sensing fluorescent compound derived from cyanoacrylic acid. J. Mater. Chem. **20**(2), 292–298 (2010)

46. Zhang, X., Chi, Z., Xu, B., Li, H., Yang, Z., Li, X., Liu, S., Zhang, Y., Xu, J.: Synthesis of blue light emitting bis (triphenylethylene) derivatives: a case of aggregation-induced emission enhancement. Dyes Pigm. **89**(1), 56–62 (2011)

47. He, J., Xu, B., Chen, F., Xia, H., Li, K., Ye, L., Tian, W.: Aggregation-induced emission in the crystals of 9, 10-distyrylanthracene derivatives: the essential role of restricted intramolecular torsion. J. Phys. Chem. C **113**(22), 9892–9899 (2009)

48. Zhang, X., Chi, Z., Xu, B., Chen, C., Zhou, X., Zhang, Y., Liu, S., Xu, J.: End-group effects of piezofluorochromic aggregation-induced enhanced emission compounds containing distyrylanthracene. J. Mater. Chem. **22**(35), 18505–18513 (2012)

49. Zhang, X., Ma, Z., Yang, Y., Zhang, X., Chi, Z., Liu, S., Xu, J., Jia, X., Wei, Y.: Influence of alkyl length on properties of piezofluorochromic aggregation induced emission compounds derived from 9, 10-bis [(N-alkylphenothiazin-3-yl) vinyl] anthracene. Tetrahedron **70**(4), 924–929 (2014)

50. Osakada, Y., Hanson, L., Cui, B.: Diarylethene doped biocompatible polymer dots for fluorescence switching. Chem. Commun. **48**(27), 3285–3287 (2012)

51. Wu, C., Schneider, T., Zeigler, M., Yu, J., Schiro, P.G., Burnham, D.R., McNeill, J.D., Chiu, D.T.: Bioconjugation of ultrabright semiconducting polymer dots for specific cellular targeting. J. Am. Chem. Soc. **132**(43), 15410–15417 (2010)

52. Wang, L., Dong, L., Bian, G.-R., Wang, L.-Y., Xia, T.-T., Chen, H.-Q.: Using organic nanoparticle fluorescence to determine nitrite in water. Anal. Bioanal. Chem. **382**(5), 1300–1303 (2005)

53. Grimland, J.L., Wu, C., Ramoutar, R.R., Brumaghim, J.L., McNeill, J.: Photosensitizer-doped conjugated polymer nanoparticles with high cross-sections for one-and two-photon excitation. Nanoscale **3**(4), 1451–1455 (2011)

54. Kim, H.Y., Bjorklund, T., Lim, S.-H., Bardeen, C.: Spectroscopic and photocatalytic properties of organic tetracene nanoparticles in aqueous solution. Langmuir **19**(9), 3941–3946 (2003)

55. Zhou, Y., Bian, G., Wang, L., Dong, L., Wang, L., Kan, J.: Sensitive determination of nucleic acids using organic nanoparticle fluorescence probes. Spectrochim. Acta Part A Mol. Biomol. Spectrosc. **61**(8), 1841–1845 (2005)

56. Howes, P., Green, M., Levitt, J., Suhling, K., Hughes, M.: Phospholipid encapsulated semiconducting polymer nanoparticles: their use in cell imaging and protein attachment. J. Am. Chem. Soc. **132**(11), 3989–3996 (2010)

57. Feng, X., Yang, G., Liu, L., Lv, F., Yang, Q., Wang, S., Zhu, D.: A convenient preparation of multi-spectral microparticles by bacteria-mediated assemblies of conjugated polymer nanoparticles for cell imaging and barcoding. Adv. Mater. **24**(5), 637–641 (2012)

58. Wang, J., Xu, X., Zhao, Y., Zheng, C., Li, L.: Exploring the application of conjugated polymer nanoparticles in chemical sensing: detection of free radicals by a synergy between fluorescent nanoparticles of two conjugated polymers. J. Mater. Chem. **21**(46), 18696–18703 (2011)

59. Xu, X., Chen, S., Li, L., Yu, G., Liu, Y.: Photophysical properties of polyphenylphenyl compounds in aqueous solutions and application of their nanoparticles for nucleobase sensing. J. Mater. Chem. **18**(22), 2555–2561 (2008)

60. Li, K., Liu, B.: Polymer-encapsulated organic nanoparticles for fluorescence and photoacoustic imaging. Chem. Soc. Rev. **43**(18), 6570–6597 (2014)

61. Alivisatos, A.P.: Semiconductor clusters, nanocrystals, and quantum dots. Science **271**(5251), 933 (1996)

62. Peng, X., Schlamp, M.C., Kadavanich, A.V., Alivisatos, A.: Epitaxial growth of highly luminescent CdSe/CdS core/shell nanocrystals with photostability and electronic accessibility. J. Am. Chem. Soc. **119**(30), 7019–7029 (1997)

63. Kasai, H., Kamatani, H., Okada, S., Oikawa, H., Matsuda, H., Nakanishi, H.: Size-dependent colors and luminescences of organic microcrystals. Jpn. J. Appl. Phys. **35**(2B), L221 (1996)

64. Kasai, H., Kamatani, H., Yoshikawa, Y., Okada, S., Oikawa, H., Watanabe, A., Itoh, O., Nakanishi, H.: Crystal size dependence of emission from perylene microcrystals. Chem. Lett. **11**, 1181–1182 (1997)

65. Fu, H., Loo, B., Xiao, D., Xie, R., Ji, X., Yao, J., Zhang, B., Zhang, L.: Multiple emissions from 1, 3-diphenyl-5-pyrenyl-2-pyrazoline nanoparticles: evolution from molecular to nanoscale to bulk materials. Angew. Chem. Int. Ed. **41**(6), 962–965 (2002)

66. Chang, C.-W., Bhongale, C.J., Lee, C.-S., Huang, W.-K., Hsu, C.-S., Diau, E.W.-G.: Relaxation dynamics and structural characterization of organic nanobelts with aggregation-induced emission. J. Phys. Chem. C **116**(28), 15146–15154 (2012)

67. Dou, C., Chen, D., Iqbal, J., Yuan, Y., Zhang, H., Wang, Y.: Multistimuli-responsive benzothiadiazole-cored phenylene vinylene derivative with nanoassembly properties. Langmuir **27**(10), 6323–6329 (2011)

68. Tong, H., Hong, Y., Dong, Y., Häußler, M., Lam, J.W., Li, Z., Guo, Z., Guo, Z., Tang, B.Z.: Fluorescent "light-up" bioprobes based on tetraphenylethylene derivatives with aggregation-induced emission characteristics. Chem. Commun. **35**, 3705–3707 (2006)

69. Bhongale, C.J., Chang, C.-W., Diau, E.W.-G., Hsu, C.-S., Dong, Y., Tang, B.-Z.: Formation of nanostructures of hexaphenylsilole with enhanced color-tunable emissions. Chem. Phys. Lett. **419**(4), 444–449 (2006)

70. Bhongale, C.J., Hsu, C.S.: Emission enhancement by formation of aggregates in hybrid chromophoric surfactant amphiphile/silica nanocomposites. Angew. Chem. Int. Ed. **45**(9), 1404–1408 (2006)

71. An, B.-K., Lee, D.-S., Lee, J.-S., Park, Y.-S., Song, H.-S., Park, S.Y.: Strongly fluorescent organogel system comprising fibrillar self-assembly of a trifluoromethyl-based cyanostilbene derivative. J. Am. Chem. Soc. **126**(33), 10232–10233 (2004)

72. Yoon, S.-J., Chung, J.W., Gierschner, J., Kim, K.S., Choi, M.-G., Kim, D., Park, S.Y.: Multistimuli two-color luminescence switching via different slip-stacking of highly fluorescent molecular sheets. J. Am. Chem. Soc. **132**(39), 13675–13683 (2010)

73. Zhang, H., Xu, X., Ji, H.-F.: Excitation-wavelength-dependent photoluminescence of a pyromellitic diimide nanowire network. Chem. Commun. **46**(11), 1917–1919 (2010)

74. Sun, Y.-P., Zhou, B., Lin, Y., Wang, W., Fernando, K.S., Pathak, P., Meziani, M.J., Harruff, B.A., Wang, X., Wang, H.: Quantum-sized carbon dots for bright and colorful photoluminescence. J. Am. Chem. Soc. **128**(24), 7756–7757 (2006)

75. Lee, K.-M., Cheng, W.-Y., Chen, C.-Y., Shyue, J.-J., Nieh, C.-C., Chou, C.-F., Lee, J.-R., Lee, Y.-Y., Cheng, C.-Y., Chang, S.Y.: Excitation-dependent visible fluorescence in decameric nanoparticles with monoacylglycerol cluster chromophores. Nat. Commun. **4**, 1544 (2013)

76. Irimpan, L., Krishnan, B., Deepthy, A., Nampoori, V., Radhakrishnan, P.: Excitation wavelength dependent fluorescence behaviour of nano colloids of ZnO. J. Phys. D: Appl. Phys. **40**(18), 5670 (2007)

77. Jang, K.: Excitation-dependent emissive properties of silicate phosphor for light converted LEDs. J. Korean Phys. Soc. **55**, 1587 (2009)

78. Huang, Y.M., Zhai, B.-G., Zhou, F.-F.: Correlation of excitation-wavelength dependent photoluminescence with the fractal microstructures of porous silicon. Appl. Surf. Sci. **254**(13), 4139–4143 (2008)

79. Ozasa, K., Nemoto, S., Maeda, M., Hara, M.: Excitation-wavelength-dependent photoluminescence evolution of CdSe/ZnS nanoparticles. J. Appl. Phys. **101**(10), 103503 (2007)

80. Mochalin, V.N., Gogotsi, Y.: Wet chemistry route to hydrophobic blue fluorescent nanodiamond. J. Am. Chem. Soc. **131**(13), 4594–4595 (2009)

81. Kasha, M.: Characterization of electronic transitions in complex molecules. Discuss. Faraday Soc. **9**, 14–19 (1950)

82. Xu, X., Xu, C., Shi, Z., Yang, C., Yu, B., Hu, J.: Identification of visible emission from ZnO quantum dots: excitation-dependence and size-dependence. J. Appl. Phys. 111(8), 083521 (2012)

83. Li, G., Zhang, Y., Wu, Y., Zhang, L.: Wavelength dependent photoluminescence of anodic alumina membranes. J. Phys: Condens. Matter **15**(49), 8663 (2003)

84. Kasai, H., Yoshikawa, Y., Seko, T., Okada, S., Oikawa, H., Mastuda, H., Watanabe, A., Ito, O., Toyotama, H., Nakanishi, H.: Optical properties of perylene microcrystals. Mol. Cryst. Liq. Cryst. **294**(1), 173–176 (1997)

85. Birks, J.: Excimers and exciplexes. Nature **214**, 1187–1190 (1967)

86. Davis, R., Saleesh Kumar, N., Abraham, S., Suresh, C., Rath, N.P., Tamaoki, N., Das, S.: Molecular packing and solid-state fluorescence of alkoxy-cyano substituted diphenylbutadienes: structure of the luminescent aggregates. J. Phys. Chem. C **112**(6), 2137–2146 (2008)

87. Hu, R., Leung, N.L., Tang, B.Z.: AIE macromolecules: syntheses, structures and functionalities. Chem. Soc. Rev. **43**(13), 4494–4562 (2014)

88. Ding, D., Li, K., Liu, B., Tang, B.Z.: Bioprobes based on AIE fluorogens. Acc. Chem. Res. **46**(11), 2441–2453 (2013)

89. Hong, Y., Lam, J.W., Tang, B.Z.: Aggregation-induced emission. Chem. Soc. Rev. **40**(11), 5361–5388 (2011)

90. Zhao, Z., Deng, C., Chen, S., Lam, J.W., Qin, W., Lu, P., Wang, Z., Kwok, H.S., Ma, Y., Qiu, H.: Full emission color tuning in luminogens constructed from tetraphenylethene, benzo-2, 1, 3-thiadiazole and thiophene building blocks. Chem. Commun. **47**(31), 8847–8849 (2011)

91. Shi, J., Wu, Y., Sun, S., Tong, B., Zhi, J., Dong, Y.: Tunable fluorescence conjugated copolymers consisting of tetraphenylethylene and fluorene units: from aggregation-induced emission enhancement to dual-channel fluorescence response. J. Polym. Sci., Part A: Polym. Chem. **51**(2), 229–240 (2013)

92. Aldred, M.P., Li, C., Zhang, G.-F., Gong, W.-L., Li, A.D., Dai, Y., Ma, D., Zhu, M.-Q.: Fluorescence quenching and enhancement of vitrifiable oligofluorenes end-capped with tetraphenylethene. J. Mater. Chem. **22**(15), 7515–7528 (2012)

93. Ananthakrishnan, S.J., Varathan, E., Ravindran, E., Somanathan, N., Subramanian, V., Mandal, A.B., Sudha, J.D., Ramakrishnan, R.: A solution processable fluorene–fluorenone oligomer with aggregation induced emission enhancement. Chem. Commun. **49**(91), 10742–10744 (2013)

94. Zhao, Z., Chen, S., Shen, X., Mahtab, F., Yu, Y., Lu, P., Lam, J.W., Kwok, H.S., Tang, B.Z.: Aggregation-induced emission, self-assembly, and electroluminescence of 4, 4′-bis (1, 2, 2-triphenylvinyl) biphenyl. Chem. Commun. **46**(5), 686–688 (2010)

Chapter 2
Preparation of Fluorescent Organic Nanoparticles

Several methods have been developed from time to time to ensure facile preparation of organic nanoparticles. The sizes of nanoparticles depend on several parameters, *viz.* amphiphilicity and molecular weight of the discrete organic molecules, the initial concentrations of organic molecule containing solutions and the miscibility of organic solvents containing organic molecules with aqueous media. The commonly used methods of the synthesis of FONs are self-assembly, polymerization, nanoprecipitation and emulsion.

2.1 Self-assembly

Basically, self-assembly is the spontaneous organization of molecules into well-defined and stable structures by the involvement of weak and noncovalent forces of attraction. Self-assembly generally occurs under equilibrium thermodynamic conditions. This process is one of the most known approaches for the synthesis of molecular nanoparticles [1]. Several fluorescent organic nanoparticles have been synthesized by following self-assembly approach with minor modifications wherever possible [2].

Amphiphilic polymers self-assemble into nano-aggregates in aqueous solutions owing to their distinctive chemistry [3, 4]. Generally, the mixture of amphiphilic polymeric and that of organic emitters dissolved in a friendly solvent is rapidly added to a poor solvent in excess amount. Organic emitters in the core get encapsulated by the aggregation of the hydrophobic segments. The hydrophilic chains on the same time behave like shells to ensure the stabilization of the obtained nanoparticles. Alternatively, conjugation of organic emitters to the hydrophobic ends (side chains) of amphiphilic polymers will generate fluorescent nanoparticles wherein organic emitters remain embedded within the polymeric matrix (Fig. 2.1). Additionally, the hydrophilic ends can be suitably decked with chemical functionalities for further

© The Author(s), under exclusive license to Springer Nature Singapore Pte Ltd. 2018   9
W. A. Wani et al., *Fluorescent Organic Nanoparticles*, SpringerBriefs in Materials,
https://doi.org/10.1007/978-981-13-2655-4_2

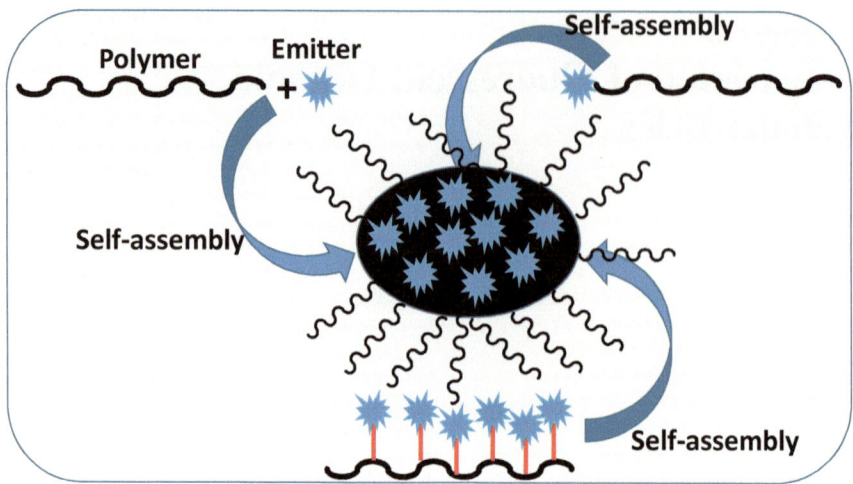

Fig. 2.1 Diagrammatic representation of polymer-encapsulated organic nanoparticle preparation *via* self-assembly

conjugation with specific targeting agents to make them eligible for multipurpose biological applications.

2.2 Polymerization

In polymerization strategies for nanoparticle preparation, the mixture of organic solvent containing both monomers and organic emitters is homogeneously disseminated into small and stable oil droplets. This is done in aqueous solution by using an emulsifier concomitantly *via* ultrasonification. Monomers are polymerized in oil droplets that generally start on the addition of initiators into the emulsion yielding organic nanoparticle dispersions (Fig. 2.2) [5, 6]. The solvent evaporation produces dispersed polymeric nanoparticles. It must be noted that emitters in this process may be either reactive or non-reactive towards monomers during polymerization [7, 8].

2.3 Emulsification

On the basis of the size of droplets, emulsions are generally classified into macro-emulsions, mini-emulsions and micro-emulsions [9]. More often, mini- and micro-emulsions are always used to form nanoparticles with small sizes (less than 500 nm). On the other hand, the droplets in macro-emulsions are sized larger than 1 mm [8]. In an emulsification process for nanoparticle synthesis, the emitters along with

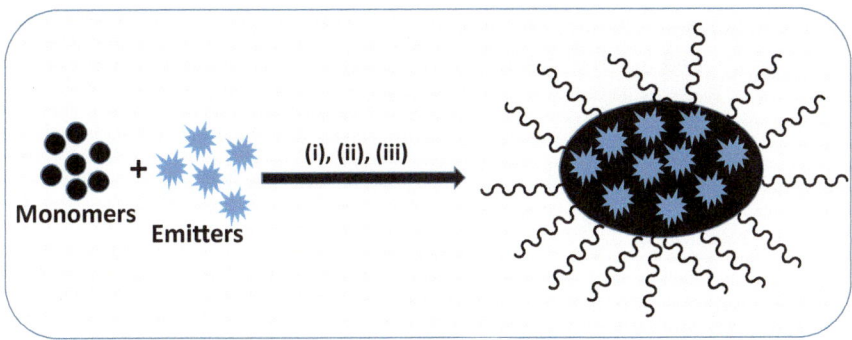

Fig. 2.2 Diagrammatic overview of *in situ* polymer-encapsulated organic nanoparticle preparation. (i) Monomers and emitters are added to water containing an emulsifier, (ii) monomers and emitters are dispersed *via* sonication and (iii) initiator is added for initiating monomer polymerization in order to yield emitter-loaded polymeric nanoparticles

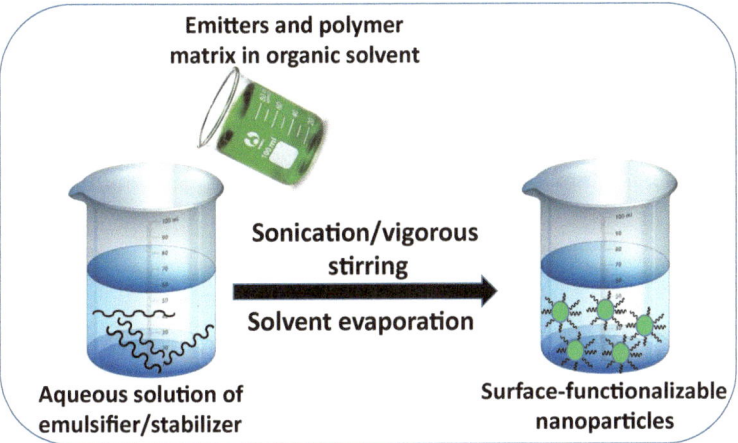

Fig. 2.3 Diagrammatic overview of polymer-encapsulated organic nanoparticle preparation from emulsion

the polymeric matrix are dissolved in an organic solvent, and a homogeneous solution is obtained. It is important to note that the organic solvent should be immiscible with water (e.g. dichloromethane). The addition of solvent into emulsifier-containing aqueous solution is done under vigorous stirring or ultrasonification. The small organic droplets formed are then stabilized by the use of an emulsifier in order to generate a homogeneous "oil-in-water" emulsion. After the evaporation of the organic solvent, stable suspensions of polymer-encapsulated nanoparticles are obtained in water. The surface of the so-produced nanoparticles can then be used for further functionalization strategies (Fig. 2.3).

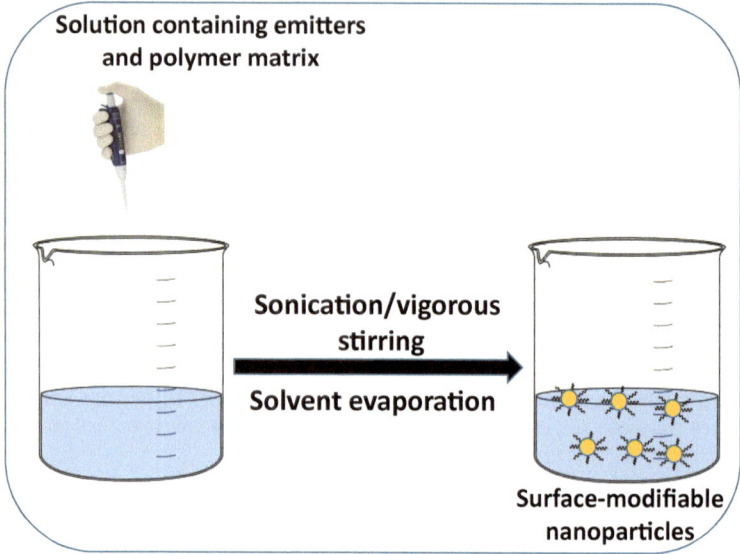

Fig. 2.4 Diagrammatic overview of the preparation of polymer-encapsulated organic nanoparticles *via* nanoprecipitation

2.4 Nanoprecipitation/Reprecipitation

Nakanishi and coworkers were the first to use nanoprecipitation approach for nanoparticle synthesis [10, 11]. This is a simple, useful and one of the most commonly used approaches for the preparation of molecular nanoparticles [1]. Nanoprecipitation significantly differs from emulsification process of nanoparticle preparation. In this process, the solvent used is always miscible with water and no emulsifier is used [12]. The solution of emitters and the encapsulation matrix is taken in an organic solvent (e.g. chloroform) and then fast added into an excess aqueous medium under ultrasonification or vigorous stirring. The instantaneous decrease in solvent hydrophobicity causes the aggregation of emitters and hydrophobic moieties of the encapsulation matrix to form nanoparticles. Besides, the orientation of the hydrophilic chains of the polymeric matrix towards water further facilitates the functionalization of the so-formed nanoparticles (Fig. 2.4). Several organic nanoparticles with fluorescence properties have been fabricated through the reprecipitation process [13–20].

A critical evaluation of the literature updates on the techniques (self-assembly, polymerization, emulsification and nanoprecipitation/reprecipitation) of organic nanoparticle preparation indicates that nanoprecipitation is the simplest and the most widely used techniques. Nanoprecipitation ensures transformation of soluble organic molecules into nanoparticles in the aqueous media and later ensures their fast screening for various analytical and biomedical applications.

References

1. Patra, A., Chandaluri, C.G., Radhakrishnan, T.: Optical materials based on molecular nanoparticles. Nanoscale **4**(2), 343–359 (2012)
2. An, B.K., Kwon, S.K., Park, S.Y.: Photopatterned arrays of fluorescent organic nanoparticles. Angew. Chem. Int. Ed. **46**(12), 1978–1982 (2007)
3. Joralemon, M.J., McRae, S., Emrick, T.: PEGylated polymers for medicine: from conjugation to self-assembled systems. Chem. Commun. **46**(9), 1377–1393 (2010)
4. Gaucher, G., Marchessault, R.H., Leroux, J.-C.: Polyester-based micelles and nanoparticles for the parenteral delivery of taxanes. J. Controlled Release **143**(1), 2–12 (2010)
5. Su, J., Chen, J., Zeng, F., Chen, Q., Wu, S., Tong, Z.: Synthesis and photochromic property of nanoparticles with spiropyran moieties via one-step miniemulsion polymerization. Polym. Bull. **61**(4), 425–434 (2008)
6. Chen, J., Zhang, P., Fang, G., Yi, P., Yu, X., Li, X., Zeng, F., Wu, S.: Synthesis and characterization of novel reversible photoswitchable fluorescent polymeric nanoparticles via one-step miniemulsion polymerization. J. Phys. Chem. B **115**(13), 3354–3362 (2011)
7. Chen, J., Zeng, F., Wu, S., Su, J., Tong, Z.: Photoreversible fluorescent modulation of nanoparticles via one-step miniemulsion polymerization. Small **5**(8), 970–978 (2009)
8. Wang, S., Kim, G., Lee, Y.-E.K., Hah, H.J., Ethirajan, M., Pandey, R.K., Kopelman, R.: Multifunctional biodegradable polyacrylamide nanocarriers for cancer theranostics—a "see and treat" strategy. ACS Nano **6**(8), 6843–6851 (2012)
9. Chen, D., Sharma, S.K., Mudhoo, A.: Handbook on Applications of Ultrasound: Sonochemistry for Sustainability. CRC press, 2011
10. Kasai, H., Nalwa, H.S., Oikawa, H., Okada, S., Matsuda, H., Minami, N., Kakuta, A., Ono, K., Mukoh, A., Nakanishi, H.: A novel preparation method of organic microcrystals. Jpn. J. Appl. Phys. 31(8A), L1132 (1992)
11. Masuhara, H., Nakanishi, H., Sasaki, K.: Single Organic Nanoparticles. Springer Science & Business Media (2003)
12. RemziáBecer, C.: Synthetic polymeric nanoparticles by nanoprecipitation. J. Mater. Chem. **19**(23), 3838–3840 (2009)
13. Jordan, A.N., Das, S., Siraj, N., de Rooy, S.L., Li, M., El-Zahab, B., Chandler, L., Baker, G.A., Warner, I.M.: Anion-controlled morphologies and spectral features of cyanine-based nanoGUMBOS—an improved photosensitizer. Nanoscale **4**(16), 5031–5038 (2012)
14. Chang, C.-W., Bhongale, C.J., Lee, C.-S., Huang, W.-K., Hsu, C.-S., Diau, E.W.-G.: Relaxation dynamics and structural characterization of organic nanobelts with aggregation-induced emission. J. Phys. Chem. C **116**(28), 15146–15154 (2012)
15. Bhardwaj, V.K., Sharma, H., Kaur, N., Singh, N.: Fluorescent organic nanoparticles (FONs) of rhodamine-appended dipodal derivative: highly sensitive fluorescent sensor for the detection of Hg^{2+} in aqueous media. New J. Chem. **37**(12), 4192–4198 (2013)
16. Wang, J., Xu, X., Shi, L., Li, L.: Fluorescent organic nanoparticles based on branched small molecule: preparation and ion detection in lithium-ion battery. ACS Appl. Mater. Interfaces **5**(8), 3392–3400 (2013)
17. Zhao, L.R., Shen, P., Zhan, X.Q., Yan, M.X., Yang, C.Y.: A highly selective chemosensor probe for Ag+ in 50% THF/H_2O. Adv. Mater. Res. 7–9 (2013) (Trans Tech Publ.)
18. Yan, H., Li, H.: Urea type of fluorescent organic nanoparticles with high specificity for $HCO_3{}^-$ anions. Sens. Actuators B: Chem **148**(1), 81–86 (2010)
19. Palayangoda, S.S., Cai, X., Adhikari, R.M., Neckers, D.C.: Carbazole-based donor-acceptor compounds: highly fluorescent organic nanoparticles. Org. Lett. **10**(2), 281–284 (2008)
20. Parthasarathy, V., Fery-Forgues, S., Campioli, E., Recher, G., Terenziani, F., Blanchard-Desce, M.: Dipolar versus octupolar triphenylamine-based fluorescent organic nanoparticles as brilliant one-and two-photon emitters for (bio) imaging. Small **7**(22), 3219–3229 (2011)

Chapter 3
Applications of Fluorescent Organic Nanoparticles

FONs are a very interesting class of materials with diverse analytical and biomedical applications. FONs have been highly attractive for cell imaging, chemosensing and drug delivery applications among others. Several polyethylene glycolated (PEGy-lated) AIE-based, biocompatible polydopamine, cross-linkable chitosan-based AIE dye, AIE dye-based, self-assembled π-conjugated and self-assembled amphiphilic fluorene oligomeric FONs have been studied for cell imaging applications [1–6]. FONs have also been extensively studied as sensors [7], drug delivery systems [8] and for other applications like photodynamic therapy [9] and apoptosis inducers of cancer cells [10].

3.1 Cell Imaging

Cell imaging agents are molecular species with ability to visualize cellular functions and follow-up the molecular process in living cells without disturbing them. These agents help in the diagnosis of several complicated diseases like cancer and neuro-logical and cardiovascular diseases. Thus, it is very important to design and develop biocompatible and water-soluble bioprobes for the diagnosis and treatment of vari-ous diseases [11–14]. Over the last few decades, several bioprobes, e.g. organic dyes, fluorescent proteins and fluorescent inorganic/organic nanoparticles among others, have been reported for biomedical applications [2, 15–27]. Despite the fact that some of the above-mentioned fluorescent materials have been commercialized, some of them have certain drawbacks as far as bioimaging applications are concerned [28]. As an example, most of the organic dyes have the drawbacks of hydrophobicity and water insolubility. These drawbacks hindered their use in biological systems. The incorporation of ionic moieties into the structures of organic dyes was tried for improving their water dispersibility; however, fluorescence quenching and toxi-city increase were the negative consequences [29]. Later, fluorescent proteins with good water solubility and biocompatibility were proved as promising materials for

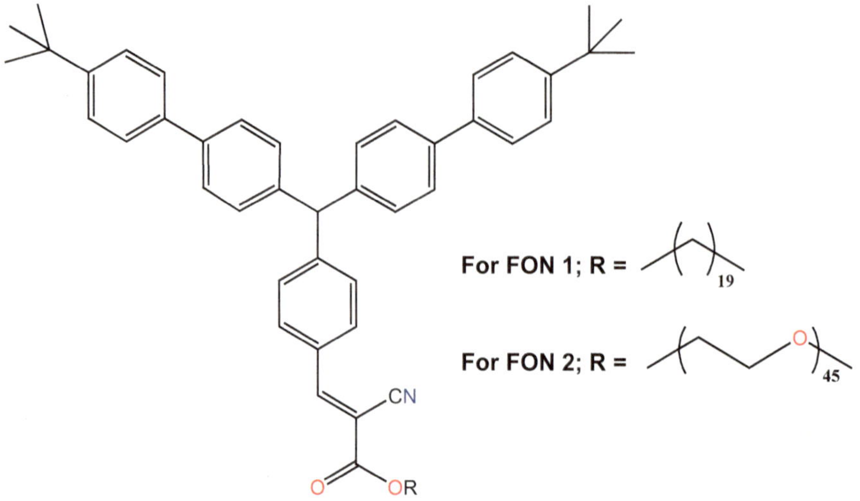

Fig. 3.1 Chemical structure of strongly solvatochromic fluorophores **1** and **2** containing long alkyl chains, 19 and 45 units long, respectively

bioimaging applications; however, certain drawbacks such as photobleaching and tough synthetic procedures hampered their development [30]. Several classes of fluorescent inorganic nanoparticles like semiconductor quantum dots, silica quantum dots, metal nanoclusters, and carbon dots with facile preparation routes and amazing optical properties were supposed to overcome the above-stated limitations of organic dyes and fluorescent proteins [30–33]. Sadly, it was their accumulation in reticuloendothelial system (RES) and non-biodegradability that raised safety issues related to their long-term administration. Thus, the design and development of cell imaging agents/bioprobes that exhibit effective renal clearance along with biodegradability properties is a very important field of research [34–37]. FONs are thought as more promising agents for biomedical applications owing to their possibility for diverse designs and biodegradability properties. Faucon et al. [38] developed strongly solvatochromic fluorophores containing alkyl chains (Fig. 3.1; **1** and **2**) that self-assembled into very bright FONs in water. The alkyl chains were observed to impart strongly hydrophobic surroundings to each fluorophore, enabling distinct emission colours between FONs where the fluorophores are associated, and their disassembled state. These colour changes were suitably harnessed for the assessment of the long-term fate of FONs in cancer cells and monocytes/macrophages. The disintegration of the FONs by monocytes/macrophages made them emit orange light, which indicated the formation of micrometer green-yellowish emitting vesicles. No significant toxicity was seen in either case that suggested the FONs as valued bioimaging agents for cell tracking with low risks of deleterious accumulation and low degradation rate.

One of the most appropriate approaches for the synthesis of FONs involves the controlled self-assembly of monodisperse π-conjugated oligomers or chromophores

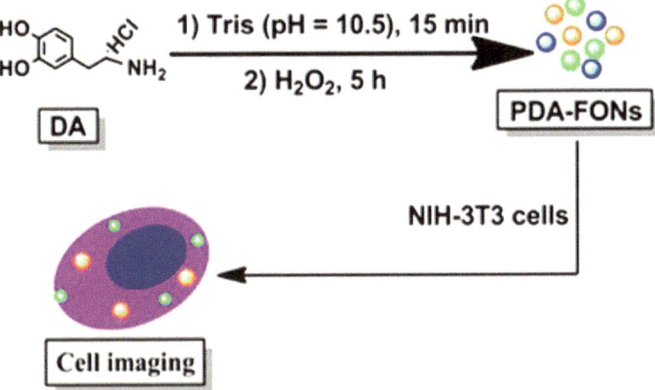

Fig. 3.2 Preparation and subsequent use of PDA-FONs as cell imaging agents in NIH-3T3 cells. Reprinted with permission from [2], Copyright © 2012, Royal Society of Chemistry

[39–41]. However, the instability in water and the strong hydrophobic interactions of FONs in aqueous media decreases their fluorescence quanta. These issues taken together drastically limit the practical biomedical applications of self-assembled structures [42]. Hence, it is the need of the hour to develop new methodologies for the synthesis of FONs that can overcome these defects [43, 44]. Taking into consideration the all-important issues of biocompatibility, Zhang et al. [2] reported cell imaging properties of water-soluble and biocompatible FONs from polydopamine (PDA-FONs) (Fig. 3.2) in mouse embryo fibroblast murine fibroblasts (NIH-3T3). The cell-internalized PDA-FONs emitted fluoresced green and green-yellow wavelengths, when excited at 405 and 458 nm, respectively, by a laser. Interestingly, the fluorescent areas were found to overlap at the locations of the cells, which indicated the cellular uptake and accumulation of PDA-FONs. The PDA-FONs were efficiently translocated into the cells and located mainly in the cytoplasm.

The same research group extended their work for the development of biocompatible FONs. polyethyleneiminglurtaraldehyde (PEI-Glu) FONs (Fig. 3.3) were prepared from polyethylenimine (PEI) and glucose and investigated for cell imaging at various fluorescent wavelengths. The FONs were reported to exhibit intense fluorescence along with high water dispersibility. The FONs were highly biocompatible with adenocarcinomic human alveolar basal epithelial (A549) cells and also showed considerable pH stability. Finally, the PEI-Glu FONs were effectively taken by cells *via* endocytosis and located at cytoplasm. A critical analysis of this report indicates a bright future of these FONs for biomedical applications. This work also envisages the development of biocompatible FONs based on other carbohydrates, biomolecules and polymers with amino groups for biomedical applications [45].

The preparation of water-soluble and biodegradable FONs has fascinated researchers. This tempted Long et al. [46] to prepare AIE-active (PEG-PABA-An-CHO FONs) (Fig. 3.4). The FONs displayed amphiphilic properties with outstand-

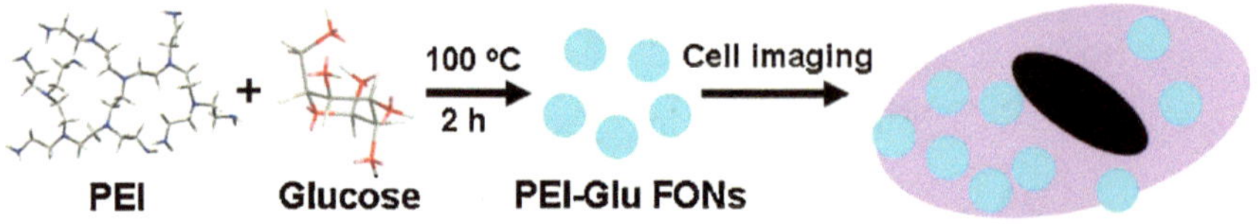

Fig. 3.3 A schematic representation of the synthesis of PEI-Glu FONs from PEI and glucose for cell imaging. Reprinted with permission from [45], Copyright © 2014 Elsevier B.V.

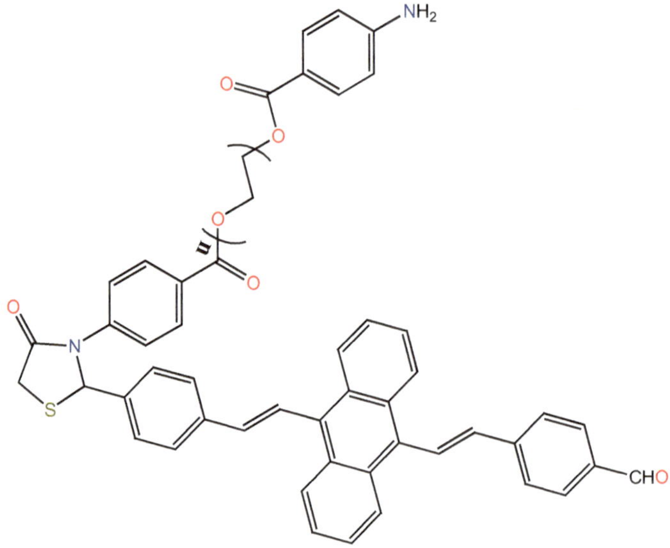

Fig. 3.4 Chemical structure of PEG-PABA-An-CHO

ing water dispersibility, desired biodegradability and negative toxicity towards cells. Besides, in aqueous solution, PEG-PABA-An-CHO FONs showed superior luminescence due to AIE. In addition, PEG-PABA-An-CHO FONs readily entered the human cervical cancer (HeLa) cells and were mainly distributed into the cytoplasm. However, PEG-PABA-An-CHO FONs could not enter into the cell nuclei because of their larger size as compared to cellular nuclear pore. A critical analysis of this report indicates that PEG-PABA-An-CHO FONs should serve as promising candidates for cell imaging.

Among the various classes of dyes, organic dipolar chromophores with π-conjugation have been quite interesting due to their ability of forming spherical nanoparticles on spontaneous aggregation in water [47, 48]. Fischer et al. [6] developed cell-permeable FONs based on self-assembled π-conjugated oligomers having high absorption cross sections and quantum yields. The nanoparticles had a tuneable density of amino groups for charge-mediated cellular uptake by a straightforward self-assembly protocol. It was shown that a single amino group per ten oligomers was sufficient for effective cellular uptake. Overall, the non-toxic nanoparticles were suit-

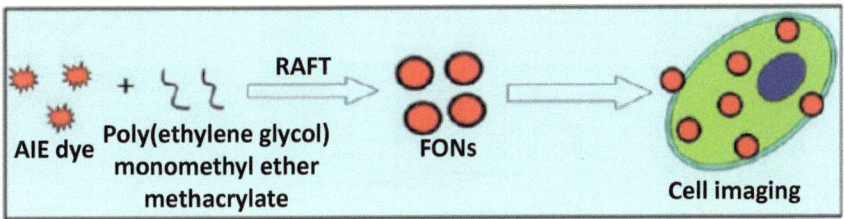

Fig. 3.5 Preparation of AIE dye-based FONs through RAFT and their cell imaging. Adapted with permission from [3], Copyright © 2013, Royal Society of Chemistry

able for both one- and two-photon cellular imaging. Later on, Amro and co-workers [49] used Suzuki-Miyaura cross-coupling protocol and efficiently prepared a dipolar chromophore with an elongated π-conjugated system that displayed a red-shifted emission. The bright orange light-emitting FONs were prepared from the naked dipole. The FONs displayed red-shifted emission along with enhanced colloidal and structural water stability and warranted bioimaging applications.

The AIE phenomenon was first reported by Tang and co-workers in 2001 [50], and since then, a number of AIE dyes based on tetraphenylethene [51–53], siloles [54, 55], cyano-substituted diarylethene [56], triphenylethene [57, 58] and distyrylanthracene [59, 60], and their derivatives have been reported for potential applications in biological and chemical sensing, opto-electronic devices and bio-imaging agents [11, 12, 61–65]. The research group of Prof. Wei [3] incorporated a cross-linkable AIE dye (Fig. 3.5) through reversible addition-fragmentation chain transfer polymerization (RAFT) into polymer nanoparticles. The nanoparticles of the dye displayed uniform size, sufficient water dispersibility, strong red fluorescence and admirable biocompatibility. These features made them promising materials for cell imaging applications. The same research group extended their work on AIE FONs using a diversity of AIE dyes such as carboxymethyl chitosan, tetraphenylethene, 9,10-distyrylanthracene, phenylalanine, diarylethene derivate dye (C18-R), cyano-substituted diarylethene derivatives, aniline, polylysine, tetraphenylethene acrylate. [1, 4, 66–73]. A critical analysis of the research work carried out by Prof. Wei's group on the development of AIE dyes as cell imaging agents indicates a promising future of such materials for use in biomedical applications. AIE dye nanoparticles have demonstrated strong fluorescence properties, excellent water dispersibility, high cellular uptake and low cytotoxicity. Besides, these nanoparticle systems provide opportunities to tune their size. Additionally, these systems contain copolymers with several types of functional groups that can react with other macromolecules and even can be changed to ensure tissue and cell-specific delivery. Thus, AIE-active polymer nanoprobes have promising future with diverse applications in nanomedicine.

The preparation of AIE FONs was carried out by other research groups as well [74–80]. Tang et al. [81] prepared 4,4′-(2,7-bis [4-phenyl]-9H-fluorene-9,9-diyl) bis (N,N,Ntrimethylbutan-1-minum) bromide (TPEFN) (Fig. 3.6), with AIEE effect. The FONs of TPEFN were formed through molecular self-aggregation by gradual

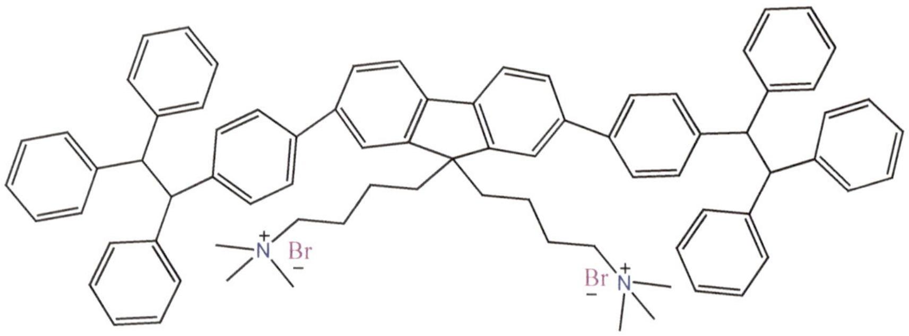

Fig. 3.6 Chemical structure of TPEN that was used for preparing nanoparticles with AIE effect

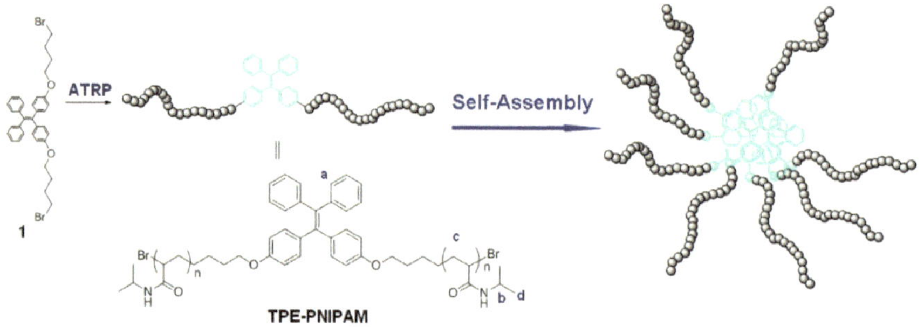

Fig. 3.7 A schematic representation of the preparation of nanoparticles from TPE-PNIPAM. Reprinted with permission from [82], Copyright (2015) American Chemical Society

increase of the water fraction in TPEFN mixed solution (methanol/water). Interestingly, a fluorescence enhancement of about 120-fold was observed upon nanoparticle formation. Addition of adenosine triphosphate (ATP) yielded larger nanoparticles of TPEFN with further fluorescence enhancement with an overall fluorescence enhancement of 420-fold with respect to the TPEFN molecular solution. Very good biocompatibility was observed with both types of nanoparticles. The FONs were efficiently internalized into cells. Overall, these FONs promise of their potential applications in cellular imaging.

Recently, Wang and co-workers [82] were successful in self-assembling temperature-sensitive organic nanoparticles based on TPE-PNIPAM (Fig. 3.7) that displayed AIE effect with strong blue-green fluorescence. Variations in temperature were used for tuning the size and fluorescence of the nanoparticles. Besides, the nanoparticles were easily internalized into HeLa. Additionally, the nanoparticles showed no cytotoxicity and also enabled tracing of the stained cells for as long as seven passages. A critical analysis of this report indicates a promising future of TPE-PNIPAM in bioimaging.

The research group of Prof. Wei has done pioneering work on FONs, and they have successfully developed FONs with promising features. In an effort to develop water

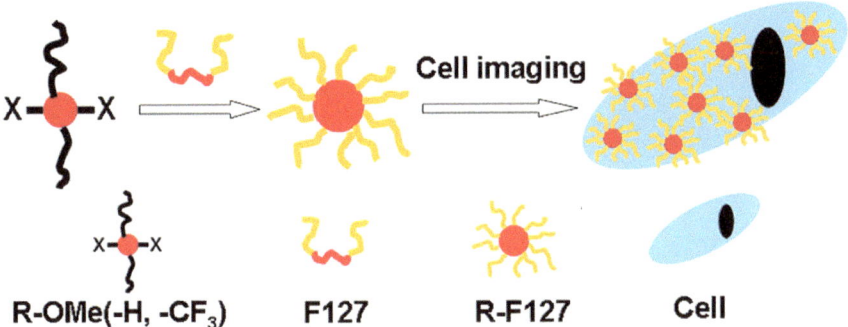

Fig. 3.8 A schematic representation of the preparation of water-soluble FONs from cyano-substituted diarylethene derivatives, and their utilization in cell imaging. Reprinted with permission from [83]. Copyright © 2014 American Chemical Society

dispersible FONs, the Wei research group prepared water-soluble red FONs (Fig. 3.8) from cyano-substituted diarylethene derivatives by using their hydrophobic inter-actions with polyoxyethylene-polyoxypropylene-polyoxyethylene triblock copoly-mer (Pluronic F127). The red FONs displayed appreciable antiaggregation-caused quenching properties in addition to broad excitation wavelengths, excellent water dispersibility and biocompatibility [83]. The same group advanced their work with some modifications, and reported red FONs based on a cyano-substituted diarylethy-lene and tetraphenylethene derivative conjugated molecule (Fig. 3.9). The FONs displayed antiaggregation-caused quenching property, high water dispersibility and excellent biocompatibility [84]. A critical analysis of these two reports indicates a promising future of FONs based on cyano-substituted diarylethene derivatives as cell imaging agents.

In an effort to develop FONs with tunable luminescence, Prof. Wei and co-workers [85] prepared FONs by hydrothermal treatment of PEI and maltose in water (Fig. 3.10). The FONs showed strong and tunable luminescence, high water dis-persibility and excellent biocompatibility. The more interesting features of the FONs were that their surface could be easily functionalized by using various functional agents like targeting agents, drugs and functional polymers. Hence, a possibility of the fabrication of multifunctional imaging agents and theranostic systems based on these FONs arises. The same group did some more work for the preparation of FONs with tunable luminescence [86]. One-pot hydrothermal treatment of starch with PEI was used to prepare FONs. The FONs displayed high water dispersibility and strong excitation-dependent fluorescence. Besides, the FONs were biocompat-ible with cells and were easily internalized within 3 h. An overall, observation of these two research reports indicates a bright future of highly water-dispersible FONs in various biomedical applications.

The tuning of the colour and photostability of FONs has been an interesting chal-lenge for the development of cell imaging agents. Besides, there is a great need to develop biodegradable substitutes of otherwise non-biodegradable quantum dots. In

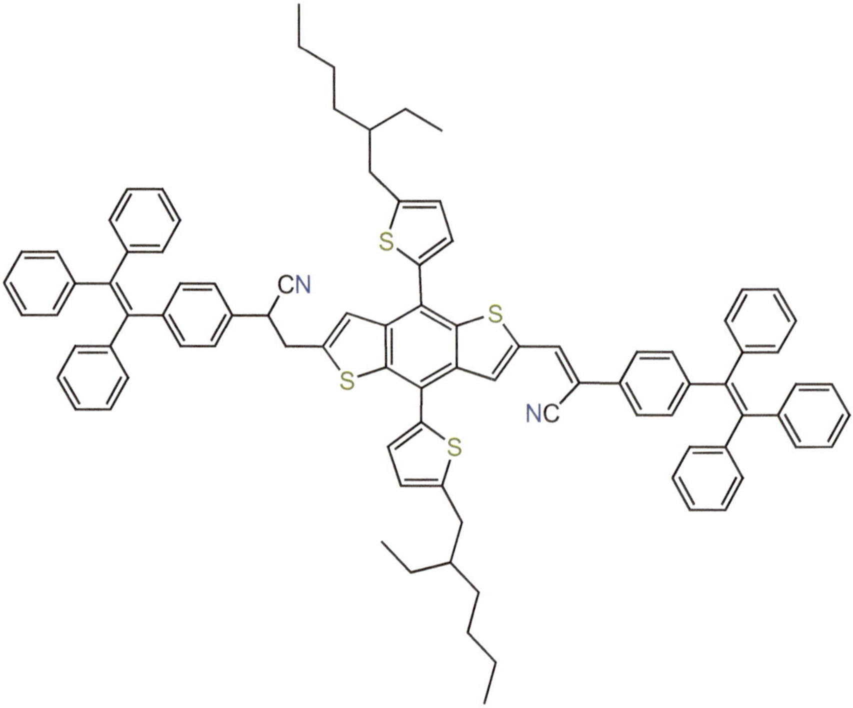

Fig. 3.9 Chemical structure of the conjugate of cyano-substituted diarylethylene and tetraphenylethene derivative

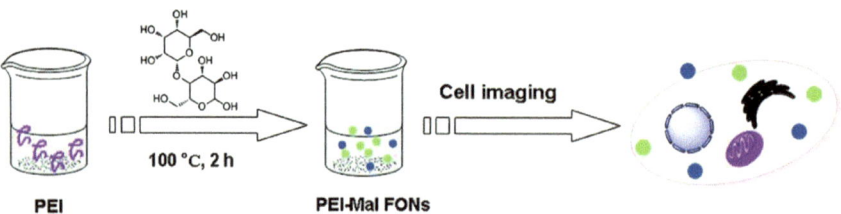

Fig. 3.10 A schematic overview of the preparation and cell imaging applications of PEI-Maltose FONs. Reprinted with permission from [85], Copyright © 2014, Royal Society of Chemistry

this direction, Trofymchuk et al. [87] encapsulated perylene diimide derivatives [one bearing bulky hydrophobic groups at the imides and the other substituted in both imide and bay regions (Lumogen Red)] (Fig. 3.11) into poly(lactic-co-glycolic acid) (PLGA). The encapsulation of the first perylene derivative resulted in aggregation and consequent emission colour change from green to red. Besides, a decrease in fluorescence quantum yield and photostability was observed. On the other hand, no aggregation was observed in Lumogen Red inside polymer nanoparticles. Additionally, a high quantum yield and photostability of Lumogen Red was reported.

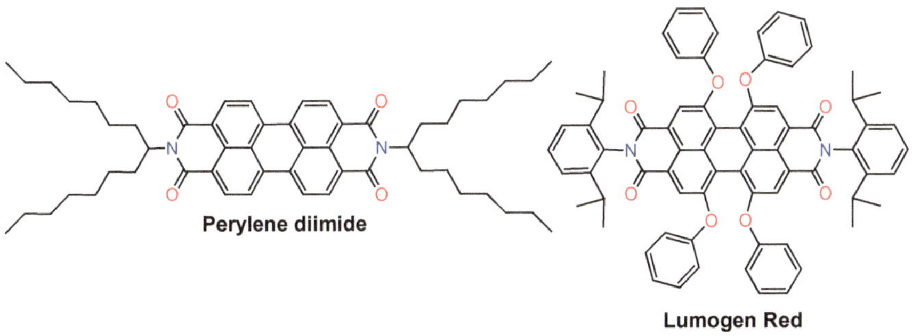

Fig. 3.11 Chemical structures of perylene diimide and Lumogen Red

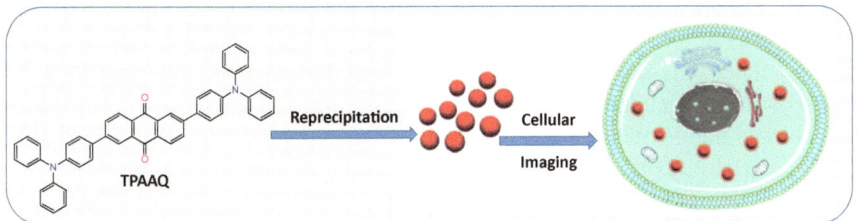

Fig. 3.12 Preparation of FONs of TPAAQ by reprecipitation and their cellular imaging. Adapted with permission from [92], Copyright © 2015, American Chemical Society

The Lumogen Red nanoparticles were more than tenfold brighter than quantum dots (QD-585). More importantly, the nanoparticles showed stability in biological media, including serum, and entered into HeLa cells spontaneously by endocytosis with no cytotoxicity. Owing to their excellent photostability, the Lumogen Red nanoparticles were thought as biodegradable substitutes of quantum dots in bioimaging.

There have been much scientific efforts for the development of near-infrared (NIR) FONs. The advantages of NIR FONs might include minimum photodamage to living cells, deep penetration into tissues and reduction of interference from the autofluorescence in the background [88–91]. Recently, Zhang et al. [92] prepared highly stable NIR FONs from an anthraquinone derivate (TPAAQ) with strong AIE (Fig. 3.12). The FONs showed high fluorescent brightness, low cytotoxicity, good pH stability and remarkable resistance against photodegradation and photobleaching. The FONs showed strong red fluorescence in A549 cells even after incubation for six generations over 15 days. It can be inferred from this research report that these NIR FONs are ideal probes for non-invasive and long-term cellular tracing and imaging.

The name cancer is associated with a huge sense of terror [93]. In terms of the number of patient deaths, cancer is ranked at second position after cardiovascular diseases [94]. Despite the availability of several classes of anticancer drugs, it is a major killer worldwide [95]. The poor diagnosis of cancer is what makes it a deadly fatal disease. Therefore, it is very important to develop methods and protocols for the diagnosis of cancer and the monitoring of tumour growth. In order to develop probes for the monitoring of cancer, Faucon and co-workers [96] developed bioconjugated FONs targeting epidermal growth factor receptor (EGFR)-overexpressing cancer cells. The authors combined a magnetic shell with an EGF ligand. The establishment of covalent binding between FONs and EGF at sub-nanomolar concentrations was confirmed from the results of dual colour fluorescence correlation spectroscopy and immunofluorescence. A strong asymmetric clustering of the EGF-conjugated FONs was seen at the membrane of human breast cancer (MDA-MB-468) cells overexpressing EGF receptors. The high recruitment of EGF-conjugated FONs was due to the EGF multivalency (4.7 EGF per FON) of the FONs, which enabled efficient EGFR activation and subsequent phosphorylation. Due to the large hydrodynamic diameter (~301 nm) of EGF-conjugated FONs, the immediate engulfment of the sequestered receptors was prevented. This provided very bright and localized spots in less than 30 min. Thus, the authors proved that the EGF-conjugated FONs may serve as ultra-bright probes of breast cancer cells with EGFR-overexpression. Recently, Xia et al. [97] developed AIE-active NIR FONs (Fig. 3.13; TPFE-Rho dots) and used them for long term in vitro cell tracking and in vivo monitoring of tumour growth. TPFE-Rho dots showed the merits of NIR fluorescent emission, large Stokes shift (~180 nm), high photostability and good biocompatibility. It was shown by in vitro cell tracing studies that TPFE-Rho dots can track human liver adenocarcinoma (SK-Hep-1) cells over 11 generations. Besides, the TPFE-Rho dots were able to monitor tumour growth for more than 19 days in a real-time manner, which was confirmed by in vivo optical imaging. A critical analysis of this report indicates that TPFE-Rho dots may be used as NIR fluorescent probes for real-time and long-term in situ and in vivo monitoring of tumour growth.

3.2 Chemosensing

Basically, a sensor detects changes or events in its environment. Originally, sensor is a transducer that may provide different forms of output, but more often uses optical or electrical signals. On stimulation by energy, a sensor alters its own state and consequently its one or more features [98]. Such changes are used to analyse the stimulant in both qualitative and quantitative limits. The more valuable changes that can be followed in this regard are the optical and photo-physical changes in a molecule. Specific transduction procedures are used by chemical sensors for providing the information about analytes. Optical absorption, luminescence, redox potential, etc., are generally the most commonly used techniques in chemical sensors [99]. Nowadays, design and development of highly sensitive and selective fluorescent probes for sensing biologically important analytes in aqueous or cellular environment is an active area of research [100–103]. Several types of fluorescent materials, such as small organic dyes [91], conjugated polymers [104, 105], organic nanoparticles [106, 107], inorganic quantum dots [108], metallic nanoclusters [109] and up-conversion nanoparticles have been used for sensing different types of analytes [110]. The chemosensing applications of FONs are discussed in the following sub-sections.

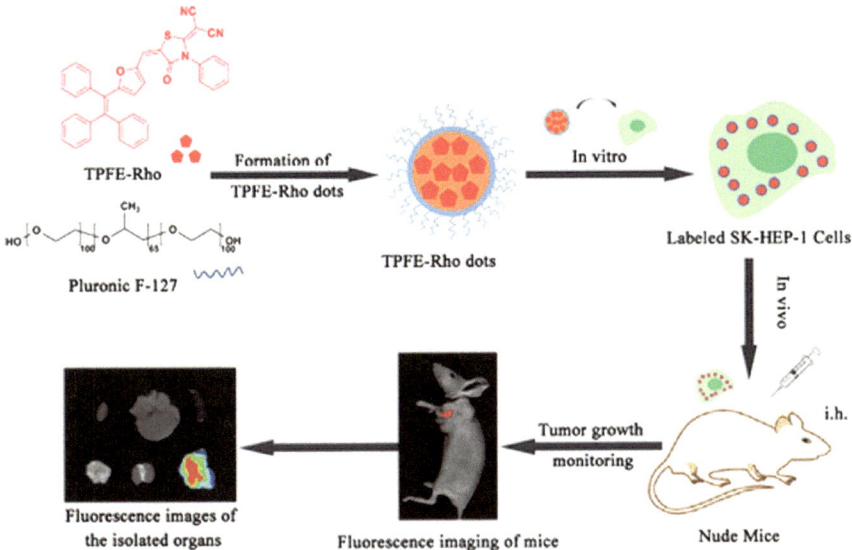

Fig. 3.13 A schematic overview of the preparation of TPFE-Rho dots for long-term monitoring of tumour growth. Reprinted with permission from [97], Copyright © 2018, American Chemical Society

3.2.1 Cation Sensing

Contamination of water resources due to chemical species in cationic states including both representative element cations and heavy metal ions poses significant risks to environmental systems and human health. Heavy metals have high density and show toxicity even at low concentrations [111]. Generally, the term heavy metals are used for the group of metals and metalloids which have atomic density greater than 4 g/cm^3 [112]. The common heavy metals are lead, cadmium, zinc, mercury, arsenic, silver, chromium, copper, iron and the platinum group elements. The general toxicities due to the exposure of lead, cadmium, mercury, arsenic, copper, zinc and aluminium include diarrhoea, gastrointestinal disorders, tremor, stomatitis, hemoglobinuria, paralysis, ataxia, depression, vomiting, convulsion and pneumonia [113].

The presence of aluminium ions in water is highly toxic to human beings and other aquatic forms of life. Drinking water contaminated with Al^{3+} ions has been known to cause damage to several types of human cells and tissues. Consumption of water contaminated with Al^{3+} ions has been reported to cause Alzheimer's and Parkinson's diseases and amyotrophic lateral sclerosis [114–116]. In order to minimize the effect of Al^{3+} ions on human health, its detection and removal is very important. However, the detection of aqueous Al^{3+} ions is comparatively difficult due to the poor coordination power, strong hydration ability and the lack of suitable spectroscopic transitions for Al^{3+} ions [117–119]. In this context, Huerta-Aguilar et al. [120] devel-

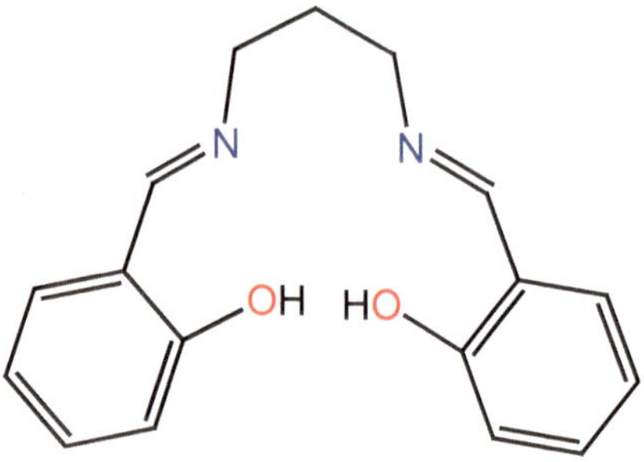

Fig. 3.14 Chemical structure of the Schiff's base, *N*,*N'*-propylenebis(salicylimine)

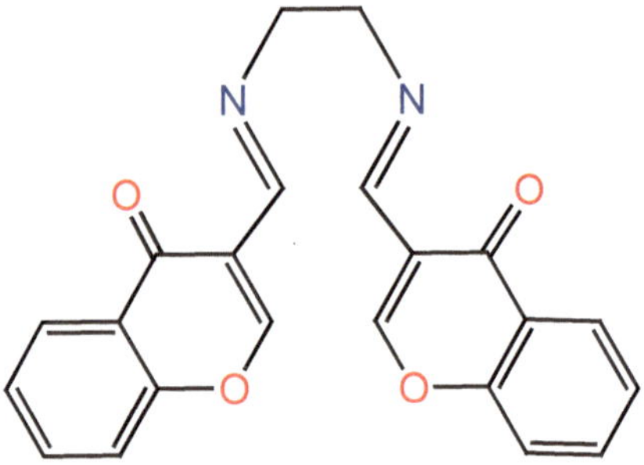

Fig. 3.15 Chemical structure of (1,2-bis[4-oxo-4H-1-benzopyran-3-ylmethylenamino]-ethane)

oped FONs of a Schiff's base, *N*,*N'*-propylenebis(salicylimine) (salpn) (Fig. 3.14) for the detection of Al^{3+} ions. Al^{3+} ions were comfortably detected with a detection limit as low as 1.24×10^{-3} mM. The fluorescence intensity increased on increasing Al^{3+} ion concentration in the presence of the salpn-organic nanoparticles. Besides, salpn-organic nanoparticles efficiently detected Al^{3+} ions out of several co-existent metal ions, such as Mn^{2+}, Mg^{2+}, Co^{2+}, Fe^{3+}, Ni^{2+}, Zn^{2+}, Sr^{2+}, Ag^+, Sm^{3+}, Al^{3+}, Cd^{2+}, Ba^{2+}, Na^+ and K^+. Additionally, the efficiency of salpn as a fluorescent probe for Al^{3+} ions was investigated in Gram-negative and Gram-positive bacteria, confirming that this chemosensor efficiently detected Al^{3+} ion in Staphylococcus aureus enclosed by a single membrane.

The sensing studies on Al^{3+} ions were also carried out by Kaur and co-workers [121], who reported excellent Al^{3+} ion chemosensing properties of FONs of (1,2-bis[4-oxo-4H-1-benzopyran-3-ylmethylenamino]-ethane) (Fig. 3.15) in an aqueous medium. The FONs displayed a detection limit of 100 nM for Al^{3+} ions, and the stoichiometry of the complex formation between the FON and Al^{3+} was 1:1.

The contamination of foodstuffs and water resources with mercury ions has been known to causes several hazards to human life. Exposure to elemental mercury causes blockage of blood vessels. It causes several serious toxic effects like damage to brain, kidneys and lungs [122]. Besides, mercury poisoning can result in several diseases such as acrodynia (pink disease) [123], Hunter-Russell syndrome [124] and Minamata disease [125]. Therefore, the detection and removal of mercuric ions from water is very important. In this direction, Singh et al. [7] reported three Biginelli-based molecules and their fluorescent nanoparticle-based chemosensors. Out of the

Fig. 3.16 Chemical structure of the chemosensor (**3**) reported by Singh et al.

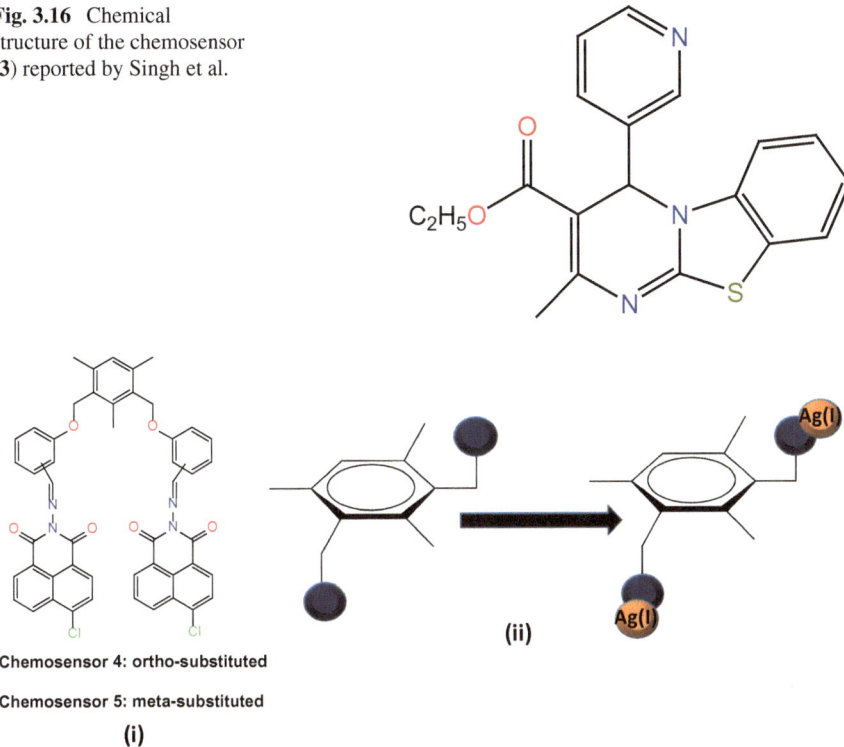

Chemosensor 4: ortho-substituted

Chemosensor 5: meta-substituted

(i)

(ii)

Fig. 3.17 (**i**) Chemical structure of the chemosensors (**4** and **5**); (**ii**) Mechanism of Ag^{+1} binding with the

three nanoparticulate chemosensors, the chemosensor **3**,(Fig. 3.16) displayed excellent sensitivity and selectivity for Hg^{2+} ions in aqueous medium. The chemosensor detected Hg^{2+} ions up to 1 nM concentrations. Additionally, it was observed that the resultant chemosensor-mercury complex detected micro-level concentrations of Cl^- ions in aqueous medium through counterion displacement mechanism.

The exposure of elemental silver or silver dust has been reported to cause Argyria or argyrosis in which skin turns blue or bluish-grey. This disease may later become generalized argyria or local argyria [126]. Thus, in order to detect silver in aqueous media, Sharma and co-workers [127] prepared FONs with an imine-linked 1, 8-naphthalimide-based dipodal chemosensor (Fig. 3.17) for the sensing of Ag^{+1} ions. The FONs were successfully used for the detection of Ag^{+1} ions in aqueous solutions in nanomolar concentrations. In addition, the chemosensor selectively sensed Ag^{+1} in a concentration range of 15–65 nM with a detection limit of 15.5 nM. The sensing experiments showed that the binding site of Ag^{+1} ions with the chemosensor consists of (–CH=N–) imine linkage along with the naphthalimide moiety (Fig. 3.17).

Caesium is a water-soluble radioactive element. Generally, it enters human body via contaminated water and gets uniformly distributed all over the body, with the

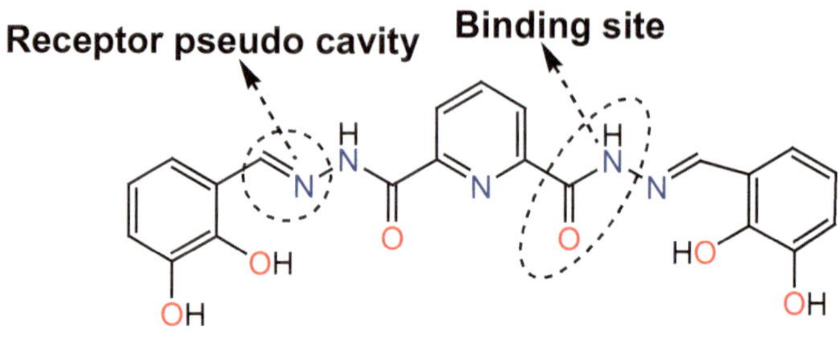

Fig. 3.18 An overview of the chemical structure of chemosensor (**6**). The encircled regions (dashed) are the imine and amide linkages that served as the receptor pseudocavity and the binding site for Cs^+ ions, respectively

maximum concentrations in soft tissues. It has been reported to cause leukaemia on long-term exposure [128], thereby making its sensing and detection a very crucial task for analytical scientists. Chopra et al. were the first to report a FON-based chemosensor for the detection of Cs^+ [129]. They prepared a nanoparticle sensor containing mixed imine and amide linkages (Fig. 3.18; **chemosensor 6**). The amide linkages defined the receptor pseudocavity, whereas the imine linkages acted as binding site and fluorescent signalling sub-unit. The nanoparticulate chemosensor was investigated for sensing in different solvents and showed selective binding for Cs^+ cations. The FONs of the chemosensor showed AIE in aqueous medium with fluorescence intensity enhancement at λ_{max} of 412 nm. Certain changes in the fluorescence emission profiles of the chemosensor were observed in low and high pH ranges; however, it was stable in 4–9 pH range. This stability made the developed chemosensor appropriate for use in environmental and biological matrices. A critical analysis of this report indicates the possibility of the use of organic dyes as sensors in aqueous media.

Water-insoluble chromium(III) species and chromium metal are not listed among the hazardous chemical agents. Nevertheless, the toxicity and carcinogenicity of chromium(VI) has been known from a long time [130]. High chromium(III) concentrations in cells have been associated with cellular DNA damage [131]. Once chromium reaches the bloodstream, it causes damage to blood cells by oxidative processes in addition to haemolysis, and kidney and liver failure. Chromium salts have also been known for their allergic reactions in certain people leading to skin ulceration [132, 133]. All these facts tempted Kaur and co-workers [134] to prepare glutathione-based tripeptide and imine-linked fluorescent sensor **7** (Fig. 3.19) for the detection of Cr^{3+} ions in aqueous medium. The FONs showed appreciable binding for Cr^{3+} ions in aqueous medium; with excited-state intramolecular proton transfer (ESIPT) featuring two emission bands, and performed as a highly sensitive and selective sensor for Cr^{3+}. Thus, such types of FONs have a bright future for sensing applications in biological and environmental matrices.

Iron poisoning has been known to cause stomach pain, nausea and bloody vomiting. Usually, the stomach pain halts after 24 h until iron passes deeper into the body,

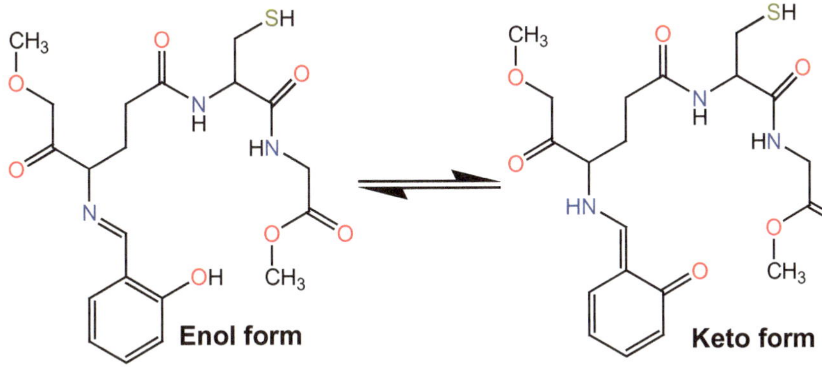

Fig. 3.19 Chemical structures of the enol and keto forms of the chemosensor (**7**)

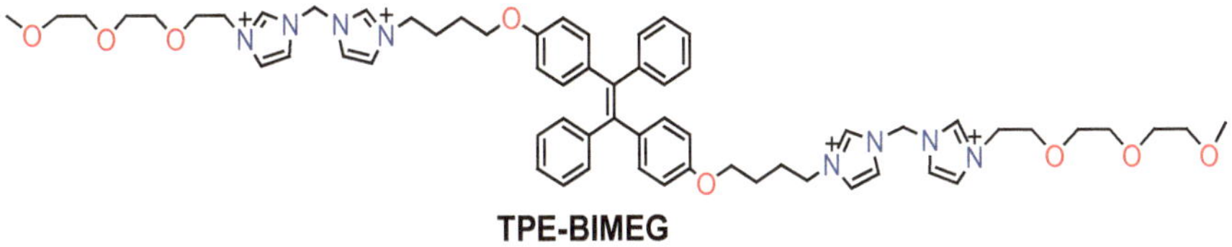

TPE-BIMEG

Fig. 3.20 Chemical structure of TPE-BIMEG

which results in metabolic acidosis. Metabolic acidosis in turn damages internal organs such as brain and liver. These facts inspired Yang and co-workers [135] to prepare FONS via self-assembly of tetraphenylethene-bis-imidazolium-oligo(ethylene glycol) (TPE-BIMEG) (Fig. 3.20) with ATP molecules in aqueous medium; for the sensing of Fe^{3+} ions. The FONs showed high selectivity and sensitivity for Fe^{3+} ions with a detection limit of 0.1 nM. Interestingly, the FONs were also capable of sensing Fe^{3+} ions in cellular environments via fluorescence quenching. This report clearly indicates the potential of organic nanoparticles based on tetraphenylethene, bis-imidazolium and oligo(ethylene glycol) for the detection and quantification of Fe^{3+} ions in cellular and environmental matrices.

Recently, Azadbakht et al. [136] reported a highly stable and fluorescent macrocyclic organic nanochemosensor, which exhibited very high selectivity and sensitivity towards Fe^{3+} ions over several other cationic species like Na^+, K^+, Cs^+, Mg^{2+}, Ca^{2+}, Ba^{2+}, Al^{3+}, Mn^{2+}, Fe^{2+}, Co^{2+}, Ni^{2+}, Cu^{2+}, Zn^{2+}, Cd^{2+}, Ag^+, Hg^{2+} and Pb^{2+} in buffered aqueous solutions. The macrocyclic chemosensor contains two naphthalene fluorophores (Fig. 3.21). A critical evaluation of these two research reports indicates a precise level of sensitivity with FONs that can be achieved while dealing with different kinds of biological and environmental matrices in the context of detection of iron contaminations.

Doubtlessly, cobalt is one of the essential trace elements for the sustenance of life. Approximately, 150–500 mg/kg is the median lethal dose (LD_{50}) for soluble

Fig. 3.21 Chemical structure of macrocyclic N_2O_2 chemosensor for the selective detection of Fe^{3+} ions

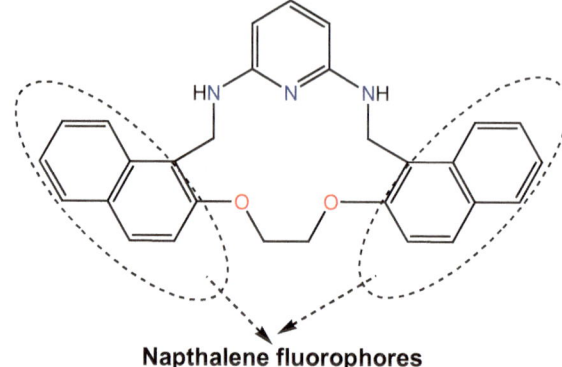

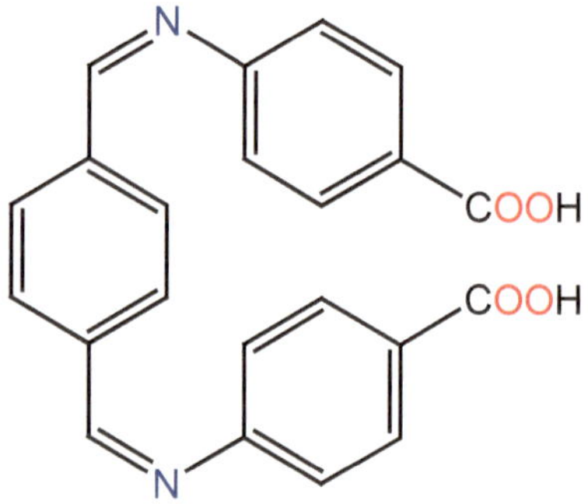

Napthalene fluorophores

Fig. 3.22 Chemical structure of 4,4′-{benzene-1,4-diylbis-[(Z)methylylidenenitrilo]}dibenzoic acid (BMBA)

cobalt salts [137]. At doses that are far less than the lethal dose, chronic ingestion of cobalt has been reported to cause serious health problems. In Canada, the addition of cobalt salts to stabilize beer foam resulted in a peculiar form of toxin-induced cardiomyopathy called beer drinker's cardiomyopathy [138, 139]. The toxic effects of cobalt include several respiratory and skin problems (contact dermatitis) [140]. Therefore, the detection of cobalt ions forms a very important task for the scientists. In this context, Mahajan and co-workers [141] documented a modest, selective and sensitive method for the detection of Co^{2+} ions using FONs of 4,4′-{benzene-1,4-diylbis-[(Z)methylylidenenitrilo]}dibenzoic acid (BMBA) (Fig. 3.22). Selective fluorescence quenching of the FONs of BMBA was observed on the addition of Co^{2+} ions, which was not affected by the presence of other co-existing ions in the matrix. A critical analysis of this finding indicates that the developed FONs of BMBA could be used in a simple way to selectively and sensitively estimate and determine Co^{2+} ions in environmental and biological samples.

It has been known that gram quantities of various copper salts produce several kinds of acute toxicities in humans. The causes of the toxicities are believed to be due to redox cycling and generation of reactive oxygen species that exhibit DNA damaging effects [142, 143]. Generally, humans are less prone to chronic copper toxicity due to the regulation of absorption and excretion by the transport systems.

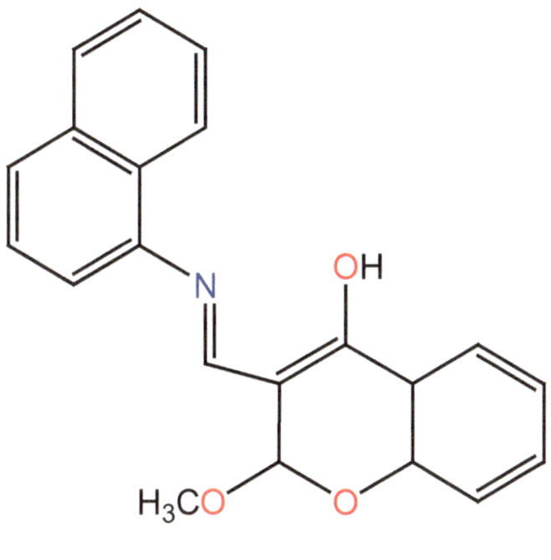

Fig. 3.23 Chemical
structure of 3-formyl
chromone

Fig. 3.23 Chemical structure of 3-formyl chromone

However, several autosomal recessive mutations in copper transport proteins can inactivate these regulatory systems, which can lead to Wilson's disease and cirrhosis of liver [132]. Additionally, increased copper levels in humans are also known to worsen the symptoms of Alzheimer's disease [144, 145]. Thus, it becomes very important to have analytical procedures for the detection of copper ions in biological and environmental matrices. In this direction, Bhardwaj et al. [146] demonstrated the use of water-soluble 3-formyl chromone-(Fig. 3.23) based FONs for the in situ sensing of Cu^{2+} ions in aqueous and biological matrices. A blue shift in absorption wavelength and efficient quenching of the emission intensity at 517 nm was observed in the presence of Cu^{2+} ions. The method demonstrates the analysis of ultra-trace amounts of Cu^{2+}, which is evident from the low limit of detection (12.3 nM) and the limit of quantification (40.59 nM). The further beauty of the nanoparticulate system was reflected from the fact that in situ formed 3-formyl chromone based FONs-Cu^{2+} assembly worked as a secondary sensor for CN^- ions. A critical analysis of this work indicates that the developed FONs can be successfully used for the detection and quantification of Cu^{2+} and CN^- ions in diverse kinds of water samples such as tap water and river water, in addition to their uses in intracellular recognition of Cu^{2+} and CN^- ions in certain cellular systems.

Lithium compounds are used in psychiatric medications. Major depressive disorder and bipolar disorders are treated by lithium compounds [147]. However, certain side effects, for example, frequent urination, shakiness of hands and increased thirst are common with the ingestion of lithium ions. Additionally, incidences of hypothyroidism, diabetes insipidus have also been observed with lithium toxicity. Thus, it becomes very important to detect lithium in biological and environmental matrices. In this direction, Kaur et al. [142] prepared FONs based on Biginelli (Fig. 3.24; **chemosensor 8**) for the selective recognition of lithium(I) ions in aqueous medium. The FONs showed selectivity towards Li (I) ions via a "fluorescence turn-on" mechanism with a limit of detection of 122 nM. Besides, the FONs were not responsive to interfering cations like Na^+, K^+, Mg^{2+} and Ca^{2+}. Additionally, the chemosensor was found to work independently of any variations in pH, which indicates its possible

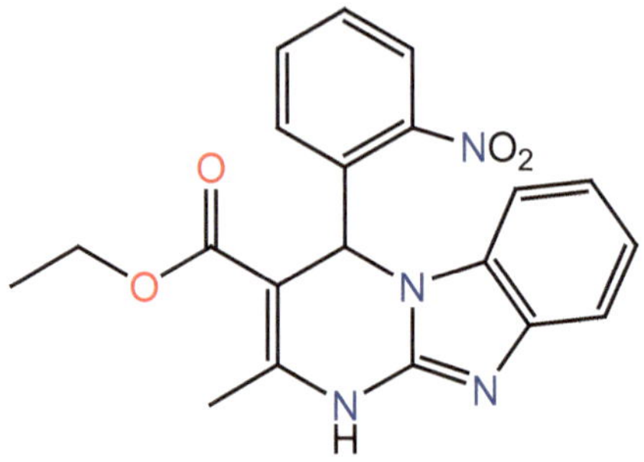

Fig. 3.24 Chemical
structure of Biginelli-based
chemosensor (**8**) used for the
selective recognition of
lithium (I) ions in aqueous
medium

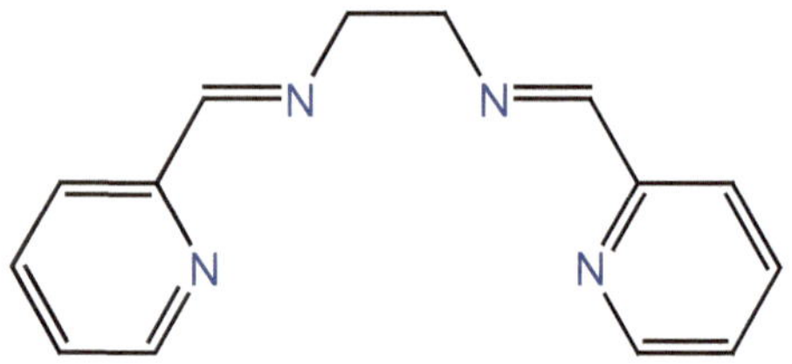

Fig. 3.25 Chemical
structure of
N,N'-bis(pyridyl-2yl-
methyl)ethylenediimine

use in biological and environmental samples of varied pH ranges for the sensing of lithium and other alkali metals.

Zinc poisoning has been associated with several risks in human beings. Nausea, vomiting, pain, cramps and diarrhoea are some of the common symptoms of zinc exposure [143]. In some cases, "zinc shakes" or "zinc chills" has also been reported due to zinc oxide inhalation [148]. In order to sense and detect zinc contamination, Huerta-Aguilar et al. [149] reported FONs of *N,N'*-bis(pyridyl-2yl-methyl)ethylenediimine (PMEDI) (Fig. 3.25) for the selective sensing and recognition of Zn^{2+} ions. The FONs detected Zn^{2+} ions at a low concentration in aqueous medium. Besides, the FONs were capable for the selective and quantitative determination of Zn^{2+} ions in multi-vitamin formulations even in presence of other nutrient metal ions. A critical analysis of this report indicates that such nanoparticulate systems can be used for checking the presence of any harmful ingredients in multivitamin formulations in addition to the checkup of the essential ingredients.

The contamination of water resources with cationic polyelectrolytes has been known to have adverse health effects on humans and aquatic life [150]. The polyelectrolytes may form undesirable by-products by reacting with other treatment chemicals added from water treatment processes like ozonation and chlorination [144]. Thus, it becomes important to design strategies for the determination of polyelectrolyte contamination in biological and environmental samples. Wu et al. [106] prepared a binary nanoparticle system based on the concept of fluorescence resonance energy transfer (FRET). The authors used poly(9,9-di-n-octylfluorenyl-2,7-diyl) (PFO; as energy donor) and poly [2-methoxy-5-(2-ethylhexyloxy)-1,4-phenylenevinylene] (MEH-PPV; as energy acceptor) for the preparation of the binary nanoparticle system. The monitoring of the FRET efficiency from PFO to MEH-PPV nanoparticles and the change in the fluorescence colours of the nanoparticle solutions were used to investigate the response of the binary nanoparticle system to cationic polyelectrolytes. Besides, the binary nanoparticle system pretreated with cationic polyelectrolyte was used to detect DNA by desorption of nanoparticles from the polyelectrolyte's chains with the detection concentration as down as 10–14 M. An analysis of this report indicates that this binary nanoparticle system has great promise in chemico-biological sensing applications.

3.2.2 Anion Sensing

The diverse and complex shapes, dependence on pH, and competition due to water in hydrogen-bonding interactions of anionic species make their host–guest chemistry very complicated [145, 151]. There are only a few reports of anion sensing in aqueous media as compared to the number of reports on cation sensing [152–154]. Therefore, design and fabrication of supra-molecules for anion quantification is a promising and challenging field of research. For the development of chemosensors for detecting anions, the binding site and the signalling subunit of the chemosensor are coupled, wherein each is destined to carry out a pre-defined function. The binding site ensures coordination with a specific anion, and the signalling unit brings changes in some photo-physical characteristics like colour or fluorescence upon anion interaction via coordination. Particularly, neutral receptors for sensing anions contain –NH fragments, which act as hydrogen-bond donors to the anion. The directional nature of the hydrogen bonds makes the receptors to distinguish between different geometries of anions; e.g. urea and thiourea act as efficient coordinating groups for anions [155, 156]. Thus, the design and development of chemosensors with high selectivity and sensitivity for the detection and monitoring of anions by using simple responses in aqueous media even with weakly coordinating anions is in high demands [145]. These demands are further enhanced by the deleterious effects of several anionic species on human health; e.g. excessive fluoride consumption leads to fluorosis [157]; hypoxia is often observed due to excessive cyanide levels, and the exposure to higher sulphide concentrations leads to several physiological and biochemical problems [158]. Hydrogen sulphate (HSO_4^-) anions are known to dissociate at high pH values. Their dissociation leads to the generation of toxic sulphate $SO4^{2-}$ anions, which cause serious irritation of skin and eyes, and even paralysis of the respiratory system [159–161]. For the sensing of HSO_4^- anions in aqueous medium, Chopra et al. [162] prepared FONs of a tripodal framework (Fig. 3.26; **9**) containing mixed donor sites for their sensing. The FONs displayed selective recognition of HSO_4^- in aqueous medium with a linear range of 0–65 mM. Besides, the FONs had the ability to detect even the lower concentrations of the analyte as the detection limit for HSO_4^- anions was 1.12 mM in a broad pH range. This report shows the potential of tripodal FONs as appropriate systems for applications in environmental or biological samples.

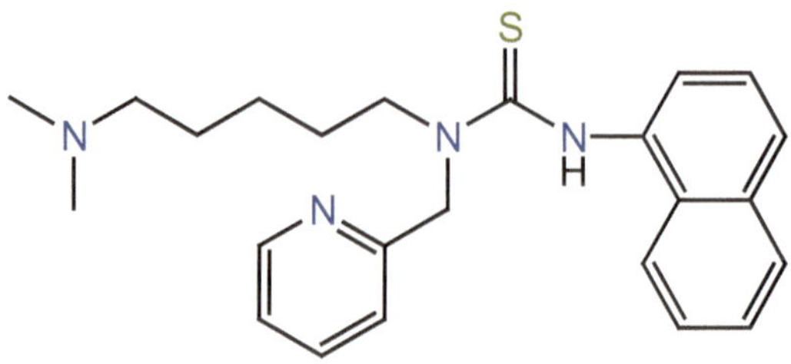

Fig. 3.26 Chemical
structure of tripodal sensor
(**9**)

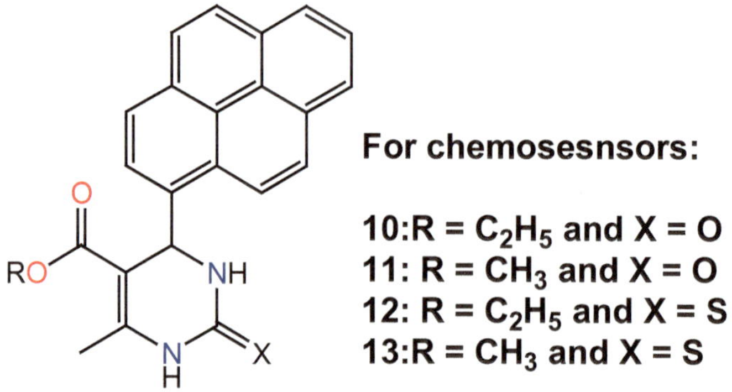

For chemosesnsors:

10:R = C$_2$H$_5$ and X = O
11: R = CH$_3$ and X = O
12: R = C$_2$H$_5$ and X = S
13:R = CH$_3$ and X = S

Fig. 3.27 Chemical structures of the chemosensors, **10–13**

Iodine is an essential mineral and is required in adequate amount by our body.
However, overdosage of iodine is toxic and may cause burning of mouth, throat and
stomach. Fever, vomiting, diarrhoea, nausea, weak pulse, cyanosis and coma are the
other symptoms of iodine overdose. Besides, the radioactive isotope of iodine is a
known carcinogen [163]. In order to sense and detect iodide anions, Kaur and co-
workers [164] prepared FONs of dihydropyrimidone derivatives (Fig. 3.27; **10–13**)
for the selective recognition of iodide anions. After adding mercury, the nanoparticles
of **12** displayed enhancement in monomer peaks of the pyrene moiety. However,
the FONs of **13** showed quenching of intensity upon Hg^{2+} addition. Interestingly,
no sensing potential was observed for the nanoparticles of **10** and **11**. In addition,
the mercury complex of **12** FONs sensed iodide anions by exhibiting quenching
in monomer and excimer emission with a detection limit of 0.2 nM in aqueous
medium. On the other hand, the mercury complex of **13** FONs showed no anion
sensing activity. Overall, the mercury complex of **12** FONs was highly sensitive and
selective towards I$^-$ anions. This nanoparticulate sensor successfully monitored the
iodide content of tap water, urine and blood serum. Thus, it may be inferred from
this report that chemosensors based on dihydropyrimidone derivatives are a novel
step towards the sensing and detection of anionic species in varied biological and
environmental conditions.

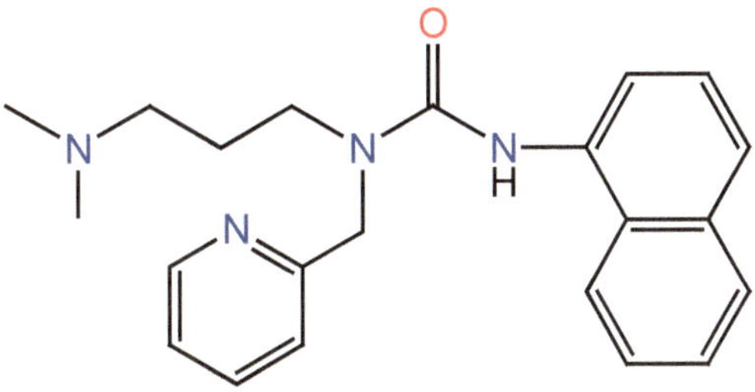

Fig. 3.28 Chemical
structure of the chemosensor
14

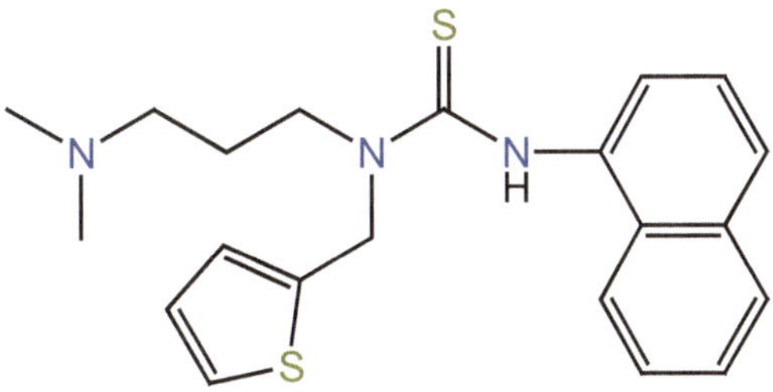

Fig. 3.29 Chemical
structure of the chemosensor
15

3.2.3 Sensing of Other Species

This sub-section highlights the development of chemosensors based on FONs for the sensing of organic and biological species in different environments. Chopra et al. [165] reported a fluorescent nanoparticulate chemosensor (Fig. 3.28; **14**) for the selective detection of spermidine. The tripodal framework of the nanoparticle-based chemosensor displayed remarkable fluorescence quenching for Fe^{3+} ions from a mixture of nineteen metal ions which was attributed to the formation of chemosensor-Fe^{3+} complex. The nanoparticulate chemosensor detected Fe^{3+} ions in as low as 1.66 mM concentration. However, the addition of spermidine increased the fluorescence intensity of the complex aqueous solution with a detection limit of 3.68 mM, which indicated the displacement of Fe^{3+} ions from the chemosensor-Fe^{3+} complex by spermidine. Interestingly, the spermidine recognition was selective and involved no interference from other biogenic amines like spermine, tyramine, 2-phenylethylamine, histamine, 1,2-diaminopropane, 1,4-diaminobutane and 1,5-diaminopentane. This report indicated the enormous potential of chemosensor-Fe^{3+} complex acts as a sensor for spermidine detection through a cation displacement assay.

The same research group [166] prepared a chemosensor based on iron complex of FONs of new receptor (Fig. 3.29; **15**) for the sensing of tyramine and 4,6-diamino-2-mercaptopyrimidine. The nanoparticulate chemosensor sensed tyramine and 4,6-diamino-2-mercaptopyrimidine with detection limits as low as 4.95 and 3.02 nM, respectively. Besides, no interference from any other biogenic amines or biothiols was observed.

Recently, the same research group did further in this direction and reported a tetrapodal receptor (Fig. 3.30), which was further processed into FONs for the

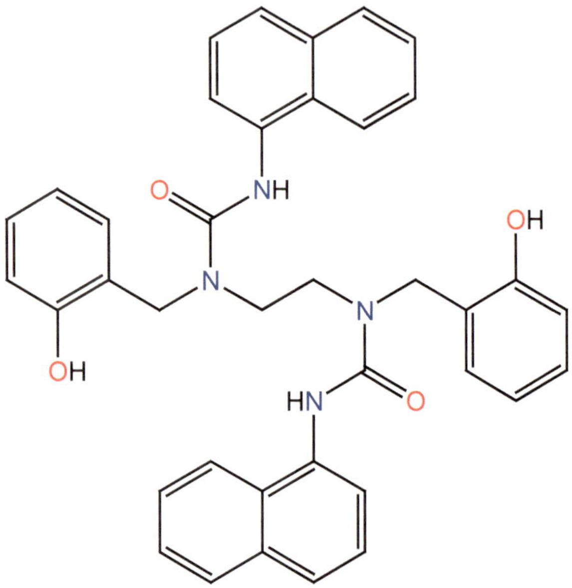

Fig. 3.30 Chemical structure of the tetrapodal receptor

determination of different biogenic amines including tyramine in aqueous medium. The processed FONs showed good affinity towards Fe^{3+} ions, and thus, the Fe^{3+}-complex of FONs was prepared and tried for the determination of different biogenic amines including tyramine. No changes in the fluorescence emission profiles of Fe^{3+}-complex of FONs was reported in the wide pH range, which indicated their potential for utility in different kinds of biological and environmental matrices. The detection of tyramine was found to be linearly proportional in response with a detection limit of 377 nM [167].

Bhardwaj et al. [168] developed FONs of dipodal rhodamine-based mercury complex (Fig. 3.31) for the selective sensing of 3-mercaptopropionic acid. The nanoparticulate chemosensor detected 3-mercaptopropionic acid in buffered aqueous medium with a detection limit of 60 nM. The overall observations were supportive of the fact that the nanoparticulate chemosensor can be of potential use for various applications in toxicology and environmental sciences.

Dopamine is an important neurotransmitter in renal, cardiovascular, central nervous and hormonal systems of the human body [169]. Dopamine can be used as a biomarker for various kinds of disorders of the central nervous system, such as Parkinson's disease, schizophrenia and anorexia [170, 171]. This makes the detection and quantification of dopamine an important practice. Recently, Ding and co-workers [172] reported FONs (Fig. 3.32; C2-F127 FONs) of Pluronic F127 that encapsulated a hydrophobic fluorescent dye C2. C2-F127 FONs had a uniform morphology with around 300 nm diameter. The FONs showed a strong orange fluorescence with an emission maximum at 561 nm. Besides, the C2-F127 FONs showed a significant quenching of fluorescence upon interaction with a dopamine. The quenching in flu-

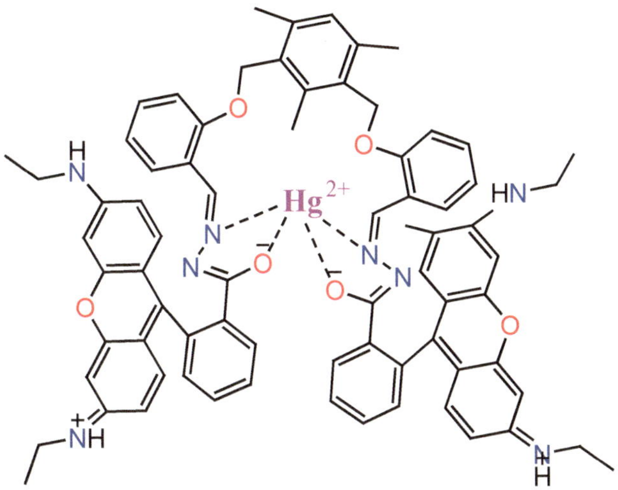

Fig. 3.31 Chemical structure of the dipodal rhodamine-based mercury complex

orescence was due to the coating of polydopamine on the surfaces of the FONs. The polydopamine coating has been assumed to bring the photo-induced charge transfer with organic dye molecules. The beauty of the FONs was that they were highly selective and sensitive to dopamine over other analytes such as uric acid, glucose, ascorbic acid, epinephrine and levodopa (l-DOPA). Owing to the appreciable dispersion and high stability of C2-F127 FONs in biological media, they were further used as label-free biosensors for the detection of dopamine in 10% serum. A remarkable sensitivity and selectivity was achieved. A critical analysis of this research report indicates that C2-F127 FONs have a bright future as sensors for the selective and sensitive detection of dopamine in biological and environmental matrices.

Penicillamine (PA) has both d- and l-enantiomeric forms (d-PA and l-PA). It is interesting to note that l-PA acts as a toxin with some toxic effects such as neuritis and osteomyelitis. However, d-PA works as a drug for the diagnosis of Wilson's disease [173]. d-PA also forms as ingredient of several medicines. Despite the important medicinal uses of d-PA, it is known to exhibit some adverse effects such as rashes on skin, oral ulceration and loss of taste [174, 175]. Thus, it is important to develop methods and protocols for the quantitative estimation and determination of d-PA in pharmaceuticals and biomedical samples. All these facts tempted Mahajan et al. [176] to prepare 1,1′-[(1E,2E)-hydrazine-1,2-diylidenedi(E)methylylidene]-dinaphthalen-2-ol (HN) (Fig. 3.33). The FONs of HN displayed red-shifted AIEE with respect to acetone solution of HN. The FONs of HN showed selective fluorescence quenching by d-PA among other analytes. Besides, the HNFONs were insensitive towards other co-existing analytes such as Cu^{2+}, Fe^{3+}, Fe^{2+}, Zn^{2+}, Ni^{2+}, Ca^{2+}, Mg^{2+}, Na^+, Al^{3+}, HCO_3^-, erythromycin, dopamine, L-penicillamine, methionine, L-cysteine and thiamine hydrochloride. HNFONs were proved to bind d-PA on account of electro-

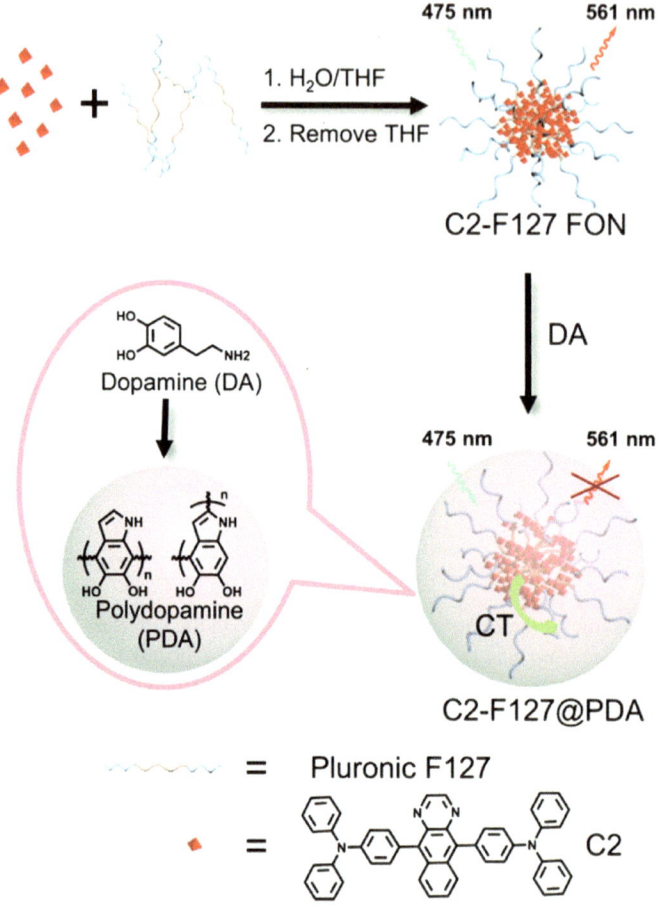

Fig. 3.32 A schematic representation of the preparation of C2-F127 FONs. Reprinted with permission from [172], Copyright © 2017, Royal Society of Chemistry

static and adsorption interactions (Fig. 3.33). Overall, the reported method showed extremely low LOD of 0.021 ppm, which is a very significant value as compared to some reported methods. Therefore, HNFONs can be successfully used for the quantitative determination of d-PA from pharmaceutical products.

Riboflavin (vitamin B2) is one of the most important biological molecules in medical sciences. Riboflavin is a water-soluble vitamin and forms an essential nutritional ingredient of human diet. Human diet if deficient in riboflavin leads to some health effects like stomatitis, inflammation of mouth corners, red tongue with sore throat, oily scaly skin rashes, itchy and watery eyes, and nasolabial folds. Riboflavin exhibits substantial fluorescence in the range of 500–600 nm with maximum wavelength at 525 nm [177]. Therefore, it is important to develop methods that can determine riboflavin in biological and environmental matrices. All these facts tempted Maha-

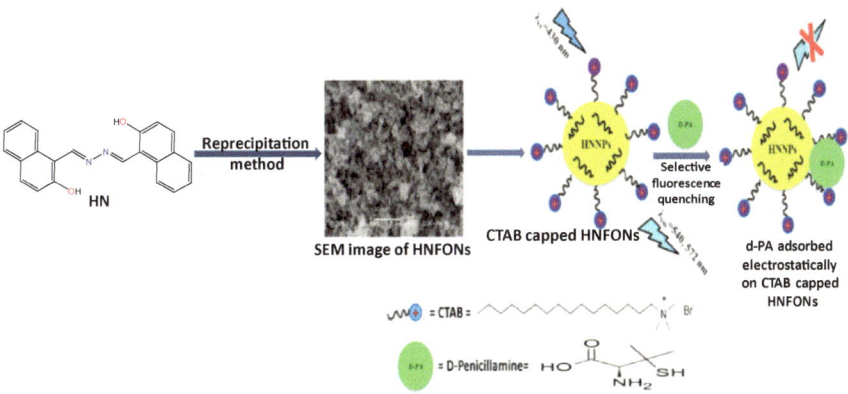

Fig. 3.33 An overview of the chemical structure of 1,1′-[(1E,2E)-hydrazine-1,2-diylidenedi(E)methylylidene]-dinaphthalen-2-ol (HN). Also shown are the SEM images of HNFONs, CTAB capped HNFONs and the mechanism of the binding of d-PA on CTAB capped HNFONs. Reprinted with permission from [176], Copyright © 2017, Springer Science Business Media New York

Fig. 3.34 Chemical structure of dianthrylidenehydrazine (AH)

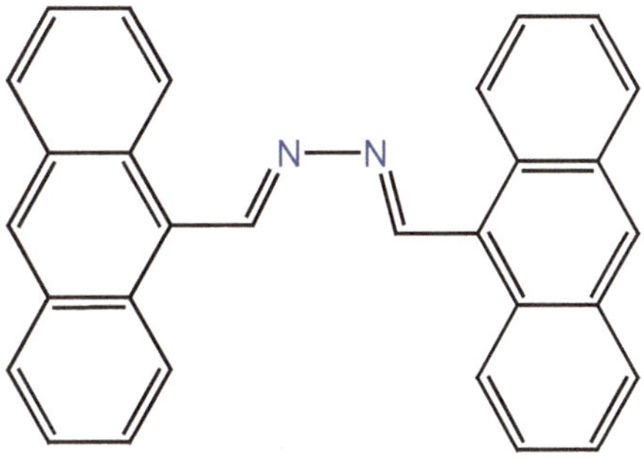

jan et al. [178] to prepare aqueous suspension of FONs of 9-anthradehdye derivative (Fig. 3.34; AH). The reported FONs displayed significant fluorescence overlap with the excitation spectrum of riboflavin, which led the authors to study the fluorescence resonance energy transfer (FRET) between FONs and riboflavin in aqueous medium. The FRET mechanism from FONs to riboflavin in the absence and presence of different concentrations of riboflavin to FONs was studied, and it was observed that the LOD for riboflavin was 0.071 µM, which is considerably low as compared to the methods available. Overall, this research report explored the potential of AHFONs as novel nanoprobes for the quantitative determination of riboflavin in biological, environmental and pharmaceutical samples.

Pseudomonas aeruginosa is an abundant Gram-negative bacterium. It colonizes in diverse natural and artificial environments [179]. Owing to its ability of survival on diverse nutritional sources and tolerance of various environmental conditions, it is counted as an emerging opportunistic human pathogen. It colonizes surfaces with moisture such as medical equipments like breathing machines, catheters. Thus, it is

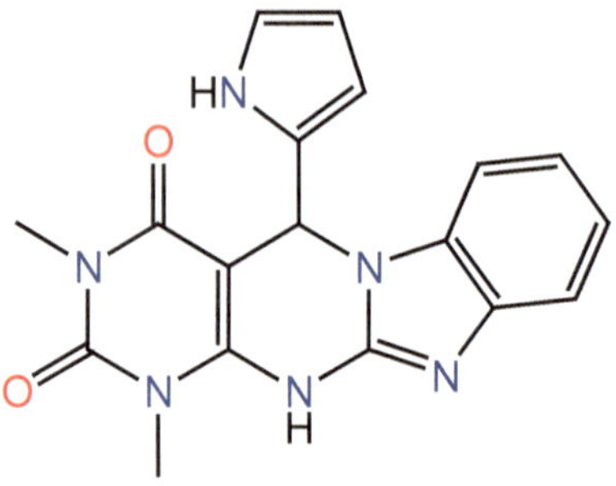

Fig. 3.35 Chemical
structure of
pyrimidine-based
chemosensor

one of the leading causes of cross infections and that too in patients hospitalized for
more than a week [180–182]. This bacterium usually causes lung infections in cystic
fibrosis patients [183, 184], affects AIDS patients who have immunocompromised
host defence mechanisms [185]. Generally, it is very tough to curb these infections,
and therefore, they represent a big therapeutic challenge mainly due to their increasing resistance against different antibiotics [186, 187]. Thus, the strategies that can
rapidly diagnose such pathogenic organisms would help the medical community for
treating various contagious diseases even before the infection becomes chronic. In
this direction, Kaur et al. [188] reported FONs of a pyrimidine-based chemosensor (Fig. 3.35) for the detection of *P. aeruginosa*. The fluorescence intensity of the
FONs selectively enhanced in the presence of *P. aeruginosa* with a detection limit of
46 colony-forming unit (CFU). Besides, the nanoparticulate sensor displayed good
antibacterial properties over a period of time. Thus, it may be envisaged from this
report that the use FONs of pyrimidine derivatives may pave way for the development
of other efficient biosensors for the sensing and detection of microbial pathogens in
different fields of biology and medicine.

3.3 Drug Delivery

Drug delivery systems (DDSs) have brought about a great revolution in the pharmaceutical field. The use of DDSs can help in avoiding some of the inherent drawbacks
of commonly used drugs, such as low solubility in physiological systems, leaching,
lower activity, unwanted interactions with different biological macromolecules other
than the target ones, toxicity, decomposition. [189, 190]. The initial studies of the
possibility of drug delivery by using FONs were done by Breton and co-workers
[191]. They synthesized some red light-emitting nanomaterials by the reprecipitation method. The nanomaterials were characterized as non-doped nanospheres with
a remarkably low polydispersity. The internalization of the reported nanoparticles
was successful in NIH-3T3 cells and displayed normal toxicity effects after 48 h of
incubation. Dual emission of the nanoparticles was observed which enabled the local-
ization of the nanoparticles within plasma membrane and cytoplasm. The potential

of self-assembled FONs formed from hexa[p-(carbonyl glycin methyl ester) phenoxy] cyclotriphosphazene (HGPCP) as potent agents for drug-loading and tracer drug delivery was realized recently [8].

Photoresponsive nanoparticles have been the preferred choice for drug delivery applications owing to their ability of controlling the release of pharmacologically important drugs via externally regulated stimulation of light [192]. More often, photoresponsive nanoparticles that are used in DDSs are constituted by a biocompatible nanocarrier and a "phototrigger". Generally, phototrigger is acted upon by a small organic molecule that helps in providing accurate control over the release of drugs and also serves as a linker molecule between nanocarrier and drug. Jana et al. [193] demonstrated the use of perylene-3-ylmethanol (Fig. 3.36) FONs as a drug delivery system. It was exciting for the authors to observe the four important functions of the perylene-3-ylmethanol nanoparticles, including acting as nanocarriers for drug molecules, phototriggers for drug release (Fig. 3.36), fluorescent chromophores for cell imaging and detectors for real-time monitoring of drug release. The perylene-3-ylmethanol nanoparticles were reported to show good cellular uptake and biocompatibility in addition to the efficient photoregulated anticancer drug release for killing HeLa cells by irradiation under visible light (≥410 nm). More importantly, the loading and unloading of the anticancer drugs was shown by the changes in the fluorescence colour of the perylene-3-ylmethanol nanoparticles.

The same research group extended their work towards the development of DDSs using FONs [194]. A nucleus-targeting photoresponsive DDS based on acridin-9-methanol (Fig. 3.37) FONs was explored for the regulated release of anticancer drug chlorambucil. The acridin-9-methanol FONs successfully acted as nucleus-targeted nanocarriers, phototrigger and as fluorescent chromophores for cell imaging. It was revealed from the in vitro experiments that the acridin-9-methanol nanoparticles of ~60 nm size were very efficient in leading to the delivery of chlorambucil to the target nucleus (Fig. 3.37), and thereby killing HeLa cells upon visible light (≥410 nm) irradiation. Overall, the targeted FONs possessed good cellular uptake property, appreciable biocompatibility along with an efficient photoregulated drug releasing ability, and therefore, are foreseen as materials of great interest in targeted intracellular controlled drug release studies.

A critical evaluation of these reports highlights the enormous potential of strongly fluorescent perylene- and acridine-based FONs as promising photoresponsive nanocarriers for use in drug delivery applications involving targeted delivery to the nuclei of cancer cells.

The same research group further extended their work and reported single component FONs based on a multi-arm PEG that was functionalized with biotin (a targeting unit) and coumarin (a fluorophore). This was used to treat HeLa cells synergistically in a site-specific and image-guided manner (Fig. 3.38). Coumarin released chlorambucil photo-controllably and also generated singlet oxygen upon UV/vis light (≥365 nm), irradiation. In addition, the polymeric organic nanoparticle containing chlorambucil (PEG-Bio-Cou-Cbl) showed approximately ~5% reduction of cell viability by a combined treatment of PDT and chemotherapy in comparison to the ~49% cell viability by PDT [195].

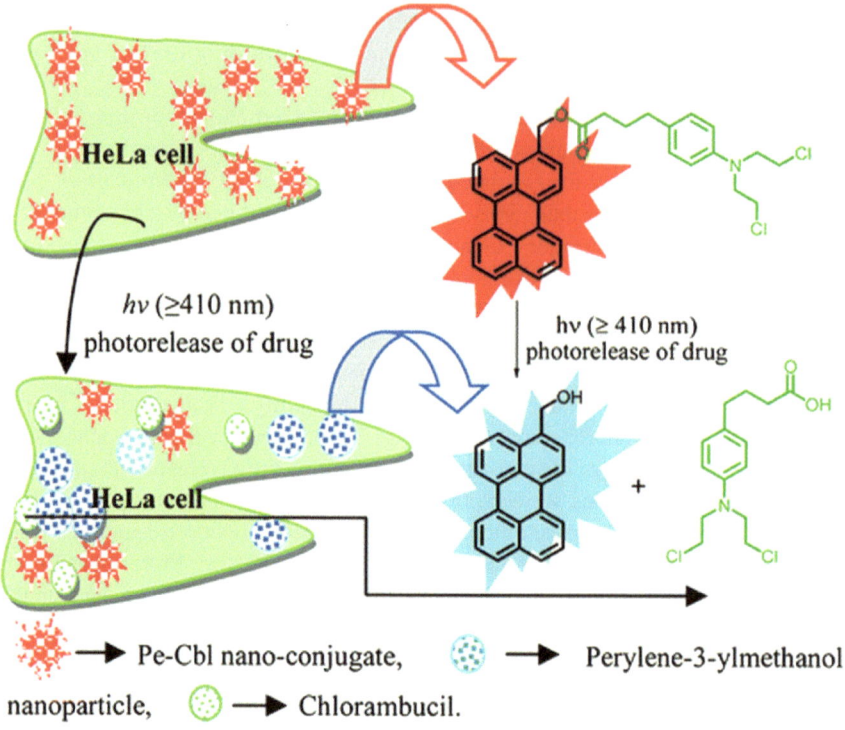

Fig. 3.36 A schematic overview of the photo-induced drug release by perylene-3-ylmethanol FONs. Reprinted with permission from [193], Copyright © 2012, American Chemical Society

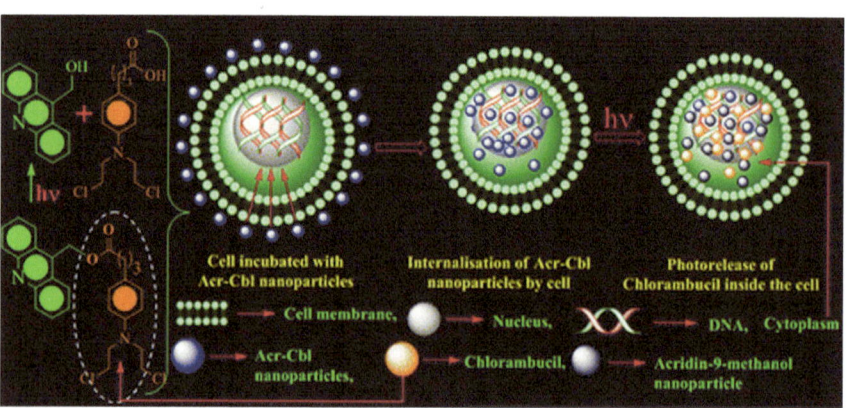

Fig. 3.37 A schematic overview of the photo-induced drug release by acridin-9-methanol FONs. Reprinted with permission from [194], Copyright © 2013, American Chemical Society

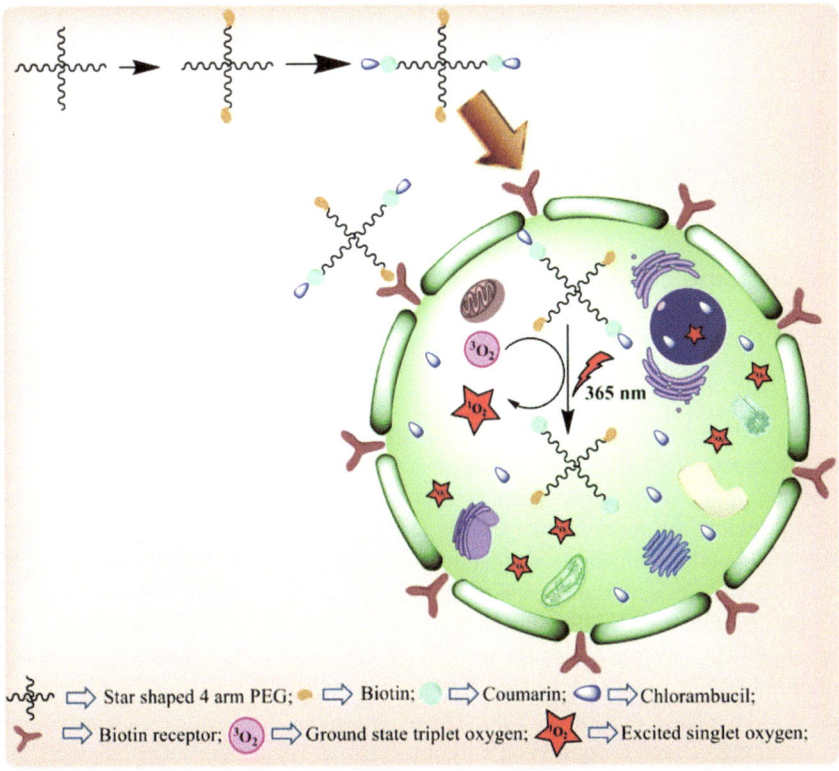

Fig. 3.38 A schematic overview of the synergic effect of PDT and chemotherapy from biotinylated coumarin FONs based on star shaped 4-arm PEG. Reprinted with permission from [195], Copyright © 2015, Royal Society of Chemistry

Recently, the same research group did a pioneering work as they reported the AIE along with an excited-state intramolecular proton transfer assisted photorelease of anticancer drugs (chlorambucil and ferulic acid) by a *p*-hydroxyphenacyl (pHP) (Fig. 3.39) phototrigger with real-time monitoring. The reported photoresponsive DDS was able to release two different drugs successively and thus may be used efficiently in combination chemotherapy. Additionally, a uniform distribution of nanoparticles of the DDS was found within the HeLa cells, which indicated their good cellular internalization. Real-time monitoring of drug release by the nanoparticulate DDS was studied through confocal microscopy wherein the cells showed yellow fluorescence due to internalization of the DDS. Both yellow and blue fluorescence was observed after 3 min of exposure to light, which indicated the partial release of the drugs. A complete change in fluorescence colour from yellow to blue was observed after 5 min of irradiation by visible light that suggested an increase in the quantity of photoreleased drugs. A critical analysis of this research report indicates that the reported FONs that are displaying excellent cellular uptake, substantial

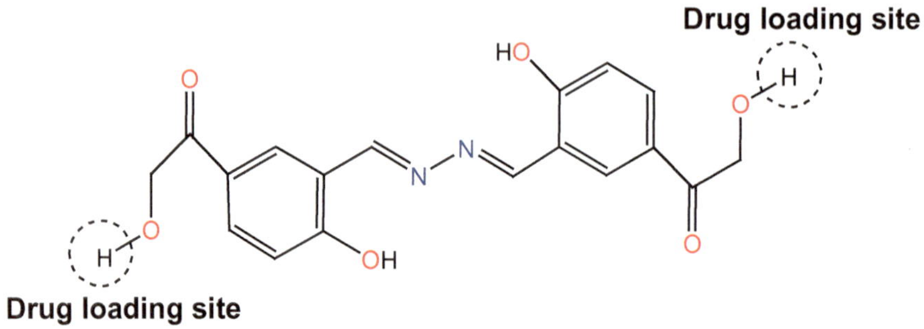

Drug loading site

Drug loading site

Fig. 3.39 Chemical structure of *p*-hydroxyphenacyl phototrigger

biocompatibility and pronounced spatiotemporal accuracy over drug release, must serve as an excellent DDS for the delivery of single and multiple drugs in future.

Microbial infections have been a curse, and the formation of bacterial biofilms on surgical devices is a major problem. Once a medical device gets infected with biofilm, the patients undergoing surgical treatment are treated with high antibiotic doses. In order to avoid the deleterious side effects due to the use of antimicrobial drugs in patients, Barman et al. [196] developed photoresponsive 1-acetylpyrene-salicylic acid (AcPy-SA) nanoparticles for the regulated release of salicylic acid, a natural antimicrobial. The phototriggering of AcPy-SA nanoparticles (Fig. 3.40) enabled the effective release of salicylic acid. The photoregulated release was dependent on the switching off and on of a visible light source. In vitro experiments showed that AcPy-SA nanoparticles efficiently delivered salicylic acid into *P. aeruginosa* cells (Fig. 3.40). The cells were potentially killed upon visible light ($\geq$410 nm) exposure. This report indicates that such photoresponsive FONs will be highly beneficial for targeted and regulated drug delivery vehicles owing to their biocompatibility, light-induced drug release property and efficient cellular uptake.

3.4 Miscellaneous Applications

This sub-sections focusses on the applications of FONs in photodynamic therapy, as apoptosis inducers, and for the study of blood–brain barrier damage. Photodynamic therapy is an effective treatment modality for cancers. This treatment procedure involves the effects of the collective presence of light, oxygen and photosensitizing drugs for achieving the required photocytotoxic effect [197–199]. Chang et al. [200] reported a series of three binary molecule conjugates. The conjugates were prepared by conjugation of a chromophore, (3,6-bis-(1-methyl-4-vinylpyridinium)-carbazole diiodide, BMVC) (Fig. 3.41) with mono-, bis- and trishydroxyl photosensitizers. BMVC plays the key roles for the recognition of cancer cells, generation of AIEE, and as Fluorescence Resonance Energy Transfer (FRET) donor. These binary con-

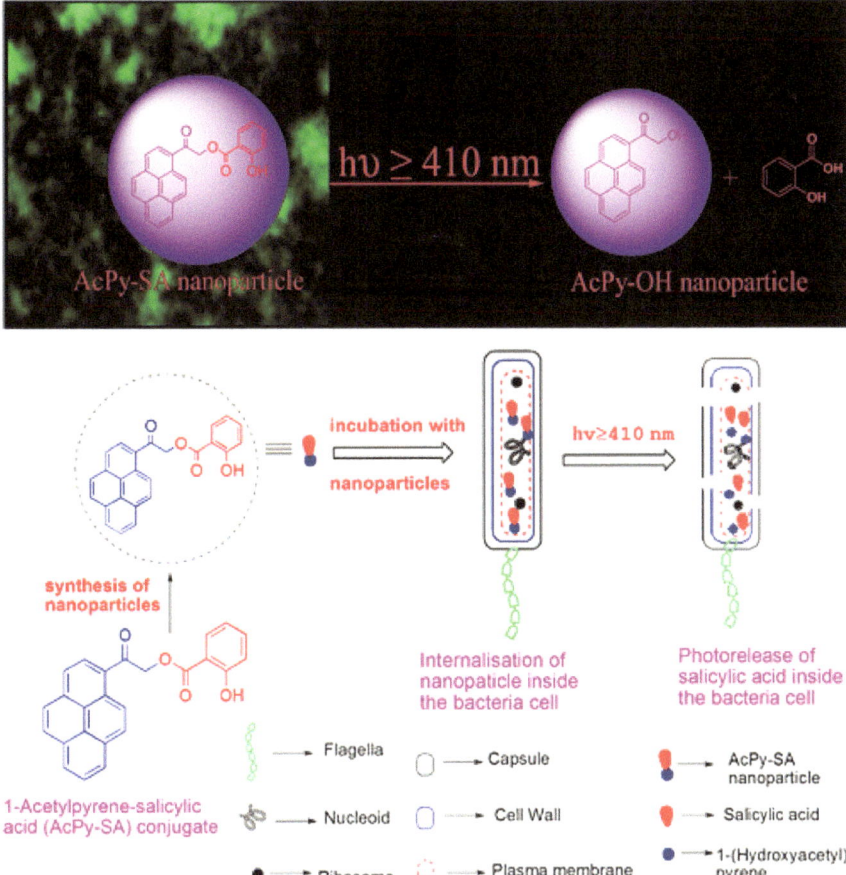

Fig. 3.40 A schematic representation of the photoresponsive drug release of salicylic acid by AcPy-SA nanoparticles. AcPy-SA nanoparticles efficiently delivered salicylic acid into *Pseudomonas aeruginosa* cells. Adapted with permission from [196], Copyright © 2014, American Chemical Society

jugates self-assembled into FONs with different degrees of AIEE. The so-formed FONs were efficient FRET and singlet oxygen generators. These FONs displayed higher phototoxicity to cancer cells than normal cells. The improved photodynamic activity was thought to be due to the aggregation of compounds in the cancer cells in the presence of light and thereby promoting PDT in cancer cells and not normal cells. The FONs also showed insignificant dark toxicity. The authors suggested that the intracellular fluorescent colour switching of the FONs upon photo-excitation may be used for cell death biomarker applications. The same group later extended their work and studied the aggregation-induced PDT enhancement due to the linear and nonlinear excited FRET of FONs. The authors designed Zn-2BPs (Fig. 3.42) for offering better FRET efficiency between BMVC and metalloporphyrin. The

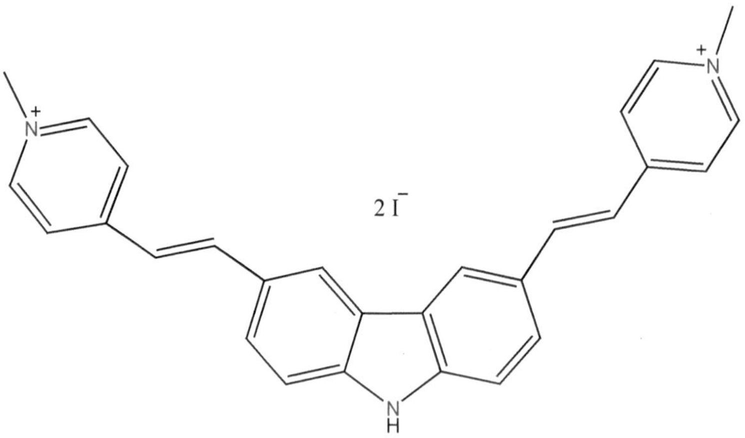

Fig. 3.41 Chemical structure of (3,6-bis-(1-methyl-4-vinylpyridinium)-carbazole diiodide

Zn-2BPs self-assembled and formed FONs that resulted in AIEE both in vitro and in the cells. Subsequently, FRET arose from the aggregated BMVC to the metalloporphyrin and additional singlet oxygen was generated for efficient PDT [9]. An observation of these two research reports indicates that the reported FONs may act as effective agents for photodynamic therapy in cancer cells and tissues.

Basically, cancer involves the uncontrolled growth and metastasis of maligned cells to virtually all over the human body. Several anticancer drugs involve apoptosis induction in proliferating cancer cells as a mechanism of their anticancer effects [201, 202]. Pramanik and co-workers [10] documented the use of fluorescent tetraethyl anthracene-9,10-diyl-9,10-bis(phosphonate) [TEABP] (Fig. 3.43) nanoparticles as selective apoptosis-inducing anticancer agents towards human histiocytic lymphoma (U937) cells. The TEABP FONs showed an IC_{50} value of 4.79 ± 4.4 μM for U937 cells after 24 h treatment. Interestingly, the FONs were selectively active against U937 cancer cells as no cytotoxicity was observed up to 30 μM concentration in normal blood and peripheral blood mononuclear cells (PBMCs). A critical observation of this report indicates that TEABP FONs might act as selective anticancer agents against U937 cells.

Blood–brain barrier (BBB) ensures stringent regulation of paracellular permeability and facilitates the entry of blood-borne substances into the brain [203]. BBB is disrupted during neurological pathologies such as ischemic stroke, which results in serious repercussions such as the increase in the concentration of vascular-derived substances into the brain [204, 205]. The size and extent of BBB damage is transient; i.e. there is an increased damage in the hyperacute phase and a reduced damage in the chronic phase [206]. The transient and widespread BBB damage forms a key to therapeutic drug development for ischemia recovery. Additionally, certain evidences have suggested that BBB injury is an important indicator of the extent of neurological damage [207, 208]. Thus, the evaluation of BBB integrity, particularly in neurolog-

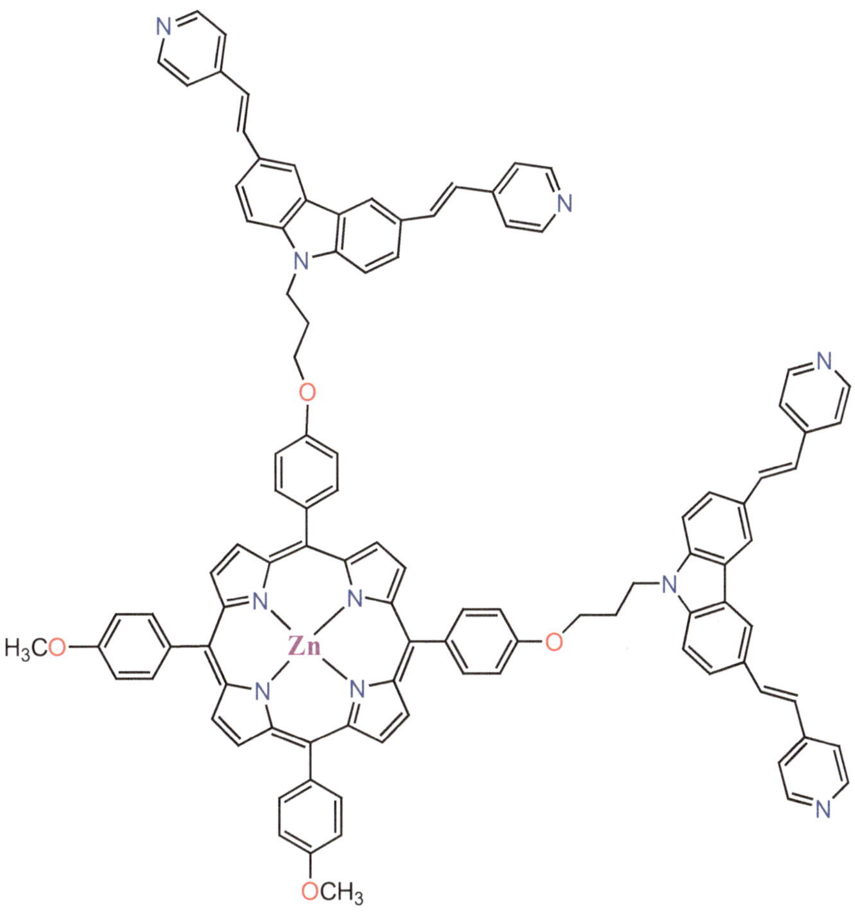

Fig. 3.42 Chemical structure of Zn-2BPS

ical pathologies for the assessment of injury and appraisal of therapeutics is very important. In this direction, Cai et al. [209] made efforts and developed a sensitive and selective optical imaging method that is based on AIEgen nanoparticles with ability to precisely test BBB integrity and also map the vascular leakage in animal models. The authors used 2,3-bis(4-(phenyl(4-(1,2,2-triphenylvinyl)phenyl)amino)-phenyl)fumaronitrile (TPETPAFN) (Fig. 3.44), which works as an AIEgen with bright red emission in the solid state. They further developed nanoparticles of TPET-PAFN with sizes of 60, 30 and 10 nm [210]. Owing to the appreciable photostability and biocompatibility of the FONs of TPETPAFN, they were in vivo investigated for BBB. BBB damage was induced in a rat model of photothrombotic ischemia (PTI). Ischemic stroke was induced precisely at a targeted single blood vessel with controlled size and well-defined boundary [211, 212]. The integrity of BBB was determined using TPETPAFN FONs during the hyperacute phase of ischemia. Attempts

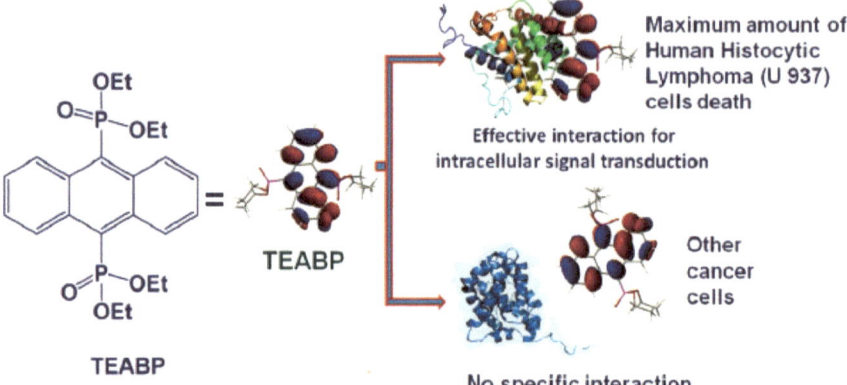

Fig. 3.43 Tetraethyl anthracene-9,10-diyl-9,10-bis(phosphonate) [TEABP] FONs and their selective apoptosis inducing anticancer effects towards U937 cells. Adapted with permission from [10], Copyright © 2013, Royal Society of Chemistry

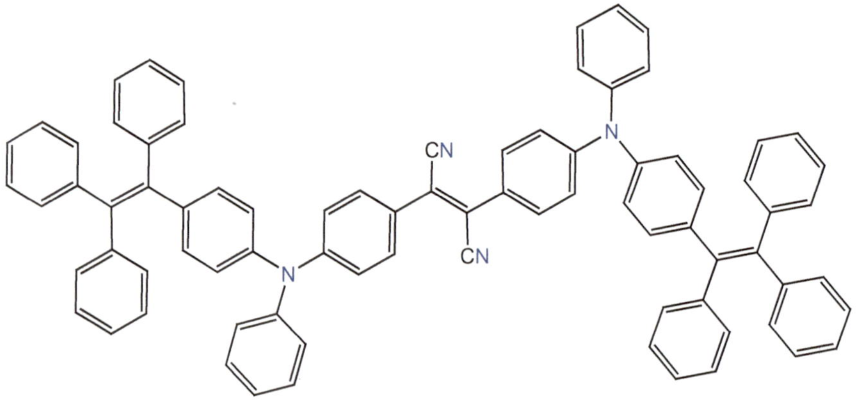

Fig. 3.44 Chemical structure of 2,3-bis (4-(phenyl(4-(1,2,2-triphenylvinyl)phenyl) amino)-phenyl)fumaronitrile (TPETPAFN)

were made to evaluate the extent of the leakage of nanoparticles in various locations of brain in response to BBB disruption after injection. It was observed that the 30 nm-sized nanoparticles were the most sensitive and selective probes for BBB damage evaluation. On the other hand, the 60 nm nanoparticles hardly passed through BBB. Interestingly, the 10 nm-sized nanoparticles showed quite poor selectivity. Overall, these results are in favour of the fact that the permeability of BBB after disruption in the PTI model is size dependent, which may be different in BBB damage induced by other neurological diseases. A critical analysis of this research report indicated that the TPETPAFN FONs can be used as excellent selective probes for the determination of BBB damage.

References

1. Zhang, X., Zhang, X., Yang, B., Wang, S., Liu, M., Zhang, Y., Tao, L., Wei, Y.: Aggregation-induced emission material based fluorescent organic nanoparticles: facile PEGylation and cell imaging applications. RSC Adv **3**(25), 9633–9636 (2013)
2. Zhang, X., Wang, S., Xu, L., Feng, L., Ji, Y., Tao, L., Li, S., Wei, Y.: Biocompatible polydopamine fluorescent organic nanoparticles: facile preparation and cell imaging. Nanoscale **4**(18), 5581–5584 (2012)
3. Zhang, X., Liu, M., Yang, B., Zhang, X., Chi, Z., Liu, S., Xu, J., Wei, Y.: Cross-linkable aggregation induced emission dye based red fluorescent organic nanoparticles and their cell imaging applications. Polym. Chem. **4**(19), 5060–5064 (2013)
4. Zhang, X., Zhang, X., Yang, B., Liu, M., Liu, W., Chen, Y., Wei, Y.: Facile fabrication and cell imaging applications of aggregation-induced emission dye-based fluorescent organic nanoparticles. Polym. Chem. **4**(16), 4317–4321 (2013)
5. Lou, X., Leenders, C.M., van Onzen, A.H., Bovee, R.A., van Dongen, J.L., Vekemans, J.A., Meijer, E.: False results caused by solvent impurity in tetrahydrofuran for MALDI TOF MS analysis of amines. J. Am. Soc. Mass Spectrom. **25**(2), 297–300 (2014)
6. Fischer, I., Petkau-Milroy, K., Dorland, Y.L., Schenning, A.P., Brunsveld, L.: Self-assembled fluorescent organic nanoparticles for live-cell imaging. Chem.—A Eur. J. **19**(49), 16646–16650 (2013)
7. Singh, A., Raj, T., Aree, T., Singh, N.: Fluorescent organic nanoparticles of Biginelli-based molecules: recognition of Hg^{2+} and Cl^- in an aqueous medium. Inorg. Chem. **52**(24), 13830–13832 (2013)
8. Li, X., Li, Z., Jing, Y., Bing, B., Li, B.: Fluorescent organic nanoparticles self-assembled from hexa [p-(carbonyl glycin methyl ester) phenoxy] cyclotriphosphazene in solution. J. Colloid Interface Sci. **375**(1), 41–49 (2012)
9. Hsieh, M.-C., Chien, C.-H., Chang, C.-C., Chang, T.-C.: Aggregation induced photodynamic therapy enhancement based on linear and nonlinear excited FRET of fluorescent organic nanoparticles. J. Mater. Chem. B **1**(18), 2350–2357 (2013)
10. Pramanik, M., Chatterjee, N., Das, S., Saha, K.D., Bhaumik, A.: Anthracene-bisphosphonate based novel fluorescent organic nanoparticles explored as apoptosis inducers of cancer cells. Chem. Commun. **49**(82), 9461–9463 (2013)
11. Feng, X., Liu, L., Wang, S., Zhu, D.: Water-soluble fluorescent conjugated polymers and their interactions with biomacromolecules for sensitive biosensors. Chem. Soc. Rev. **39**(7), 2411–2419 (2010)
12. Betzig, E., Patterson, G.H., Sougrat, R., Lindwasser, O.W., Olenych, S., Bonifacino, J.S., Davidson, M.W., Lippincott-Schwartz, J., Hess, H.F.: Imaging intracellular fluorescent proteins at nanometer resolution. Science **313**(5793), 1642–1645 (2006)
13. Dubertret, B., Skourides, P., Norris, D.J., Noireaux, V., Brivanlou, A.H., Libchaber, A.: In vivo imaging of quantum dots encapsulated in phospholipid micelles. Science **298**(5599), 1759–1762 (2002)
14. Giepmans, B.N., Adams, S.R., Ellisman, M.H., Tsien, R.Y.: The fluorescent toolbox for assessing protein location and function. Science **312**(5771), 217–224 (2006)
15. Jin, R.: Quantum sized, thiolate-protected gold nanoclusters. Nanoscale **2**(3), 343–362 (2010)
16. Hong, Y., Lam, J.W., Tang, B.Z.: Aggregation-induced emission. Chem. Soc. Rev. **40**(11), 5361–5388 (2011)
17. Wang, X., Xu, S., Xu, W.: Synthesis of highly stable fluorescent Ag nanocluster@ polymer nanoparticles in aqueous solution. Nanoscale **3**(11), 4670–4675 (2011)
18. Wu, X., He, X., Wang, K., Xie, C., Zhou, B., Qing, Z.: Ultrasmall near-infrared gold nanoclusters for tumor fluorescence imaging in vivo. Nanoscale **2**(10), 2244–2249 (2010)
19. Hui, J., Zhang, X., Zhang, Z., Wang, S., Tao, L., Wei, Y., Wang, X.: Fluoridated HAp: Ln^{3+} (Ln = Eu or Tb) nanoparticles for cell-imaging. Nanoscale **4**(22), 6967–6970 (2012)

20. Ma, C., Zhang, X., Yang, L., Wu, Y., Liu, H., Zhang, X., Wei, Y.: Preparation of fluorescent organic nanoparticles from polyethylenimine and sucrose for cell imaging. Mater. Sci. Eng., C **68**, 37–42 (2016)
21. Zhang, T., Xu, H., Wang, H., Zhu, J., Zhai, Y., Bai, X., Dong, B., Song, H.: Green fluorescent organic nanoparticles based on carbon dots and self-polymerized dopamine for cell imaging. RSC Adv. **7**(46), 28987–28993 (2017)
22. Wang, X., Liu, L., Zhu, S., Peng, J., Li, L.: Preparation of exciplex-based fluorescent organic nanoparticles and their application in cell imaging. RSC Adv. **7**(65), 40842–40848 (2017)
23. Shi, Y., Jiang, R., Liu, M., Fu, L., Zeng, G., Wan, Q., Mao, L., Deng, F., Zhang, X., Wei, Y.: Facile synthesis of polymeric fluorescent organic nanoparticles based on the self-polymerization of dopamine for biological imaging. Mater. Sci. Eng., C **77**, 972–977 (2017)
24. Shi, Y., Xu, D., Liu, M., Fu, L., Wan, Q., Mao, L., Dai, Y., Wen, Y., Zhang, X., Wei, Y.: Room temperature preparation of fluorescent starch nanoparticles from starch-dopamine conjugates and their biological applications. Mater. Sci. Eng., C **82**, 204–209 (2018)
25. Jiang, R., Liu, M., Chen, T., Huang, H., Huang, Q., Tian, J., Wen, Y., Cao, Q.-Y., Zhang, X., Wei, Y.: Facile construction and biological imaging of cross-linked fluorescent organic nanoparticles with aggregation-induced emission feature through a catalyst-free azide-alkyne click reaction. Dyes Pigm. **148**, 52–60 (2018)
26. Guo, L., Xu, D., Huang, L., Liu, M., Huang, H., Tian, J., Jiang, R., Wen, Y., Zhang, X., Wei, Y.: Facile construction of luminescent supramolecular assemblies with aggregation-induced emission feature through supramolecular polymerization and their biological imaging. Mater. Sci. Eng., C **85**, 233–238 (2018)
27. Luo, W., Jiang, R., Liu, M., Wan, Q., Tian, J., Wen, Y., Cao, Q.-Y., Hui, J., Zhang, X., Wei, Y.: Synthesis of fluorescent dendrimers with aggregation-induced emission features through a one-pot multi-component reaction and their utilization for biological imaging. J. Colloid Interface Sci. **509**, 327–333 (2018)
28. Sokolova, V., Epple, M.: Synthetic pathways to make nanoparticles fluorescent. Nanoscale **3**(5), 1957–1962 (2011)
29. Qin, A., Lam, J.W., Tang, B.Z.: Luminogenic polymers with aggregation-induced emission characteristics. Prog. Polym. Sci. **37**(1), 182–209 (2012)
30. Cai, Z., Ye, Z., Yang, X., Chang, Y., Wang, H., Liu, Y., Cao, A.: Encapsulated enhanced green fluorescence protein in silica nanoparticle for cellular imaging. Nanoscale **3**(5), 1974–1976 (2011)
31. Díez, I., Ras, R.H.: Fluorescent silver nanoclusters. Nanoscale **3**(5), 1963–1970 (2011)
32. Boisselier, E., Astruc, D.: Gold nanoparticles in nanomedicine: preparations, imaging, diagnostics, therapies and toxicity. Chem. Soc. Rev. **38**(6), 1759–1782 (2009)
33. Michalet, X., Pinaud, F., Bentolila, L., Tsay, J., Doose, S., Li, J., Sundaresan, G., Wu, A., Gambhir, S., Weiss, S.: Quantum dots for live cells, in vivo imaging, and diagnostics. Science **307**(5709), 538–544 (2005)
34. Choi, H.S., Liu, W., Misra, P., Tanaka, E., Zimmer, J.P., Ipe, B.I., Bawendi, M.G., Frangioni, J.V.: Renal clearance of quantum dots. Nat. Biotechnol. **25**(10), 1165–1170 (2007)
35. Liu, W., Choi, H.S., Zimmer, J.P., Tanaka, E., Frangioni, J.V., Bawendi, M.: Compact cysteine-coated CdSe (ZnCdS) quantum dots for in vivo applications. J. Am. Chem. Soc. **129**(47), 14530–14531 (2007)
36. Zhou, C., Long, M., Qin, Y., Sun, X., Zheng, J.: Luminescent gold nanoparticles with efficient renal clearance. Angew. Chem. **123**(14), 3226–3230 (2011)
37. Zhou, C., Hao, G., Thomas, P., Liu, J., Yu, M., Sun, S., Öz, O.K., Sun, X., Zheng, J.: Near-infrared emitting radioactive gold nanoparticles with molecular pharmacokinetics. Angew. Chem. **124**(40), 10265–10269 (2012)
38. Faucon, A., Benhelli-Mokrani, H., Córdova, L.A., Brulin, B., Heymann, D., Hulin, P., Nedellec, S., Ishow, E.: Are fluorescent organic nanoparticles relevant tools for tracking cancer cells or macrophages? Adv. Healthc. Mater. **4**(17), 2727–2734 (2015)
39. Ye, F., Wu, C., Jin, Y., Chan, Y.-H., Zhang, X., Chiu, D.T.: Ratiometric temperature sensing with semiconducting polymer dots. J. Am. Chem. Soc. **133**(21), 8146–8149 (2011)

40. Wu, C., Schneider, T., Zeigler, M., Yu, J., Schiro, P.G., Burnham, D.R., McNeill, J.D., Chiu, D.T.: Bioconjugation of ultrabright semiconducting polymer dots for specific cellular targeting. J. Am. Chem. Soc. **132**(43), 15410–15417 (2010)
41. Chan, Y.-H., Jin, Y., Wu, C., Chiu, D.T.: Copper (II) and iron (II) ion sensing with semiconducting polymer dots. Chem. Commun. **47**(10), 2820–2822 (2011)
42. Xie, R., Xiao, D., Fu, H., Ji, X., Yang, W., Yao, J.: Effect of PVA on the growth and the optical properties of perylene nanocrystals. New J. Chem. **25**(11), 1362–1364 (2001)
43. Wang, M., Zhang, G., Zhang, D., Zhu, D., Tang, B.Z.: Fluorescent bio/chemosensors based on silole and tetraphenylethene luminogens with aggregation-induced emission feature. J. Mater. Chem. **20**(10), 1858–1867 (2010)
44. Zhao, Z., Chen, S., Shen, X., Mahtab, F., Yu, Y., Lu, P., Lam, J.W., Kwok, H.S., Tang, B.Z.: Aggregation-induced emission, self-assembly, and electroluminescence of 4, 4′-bis (1, 2, 2-triphenylvinyl) biphenyl. Chem. Commun. **46**(5), 686–688 (2010)
45. Zhang, X., Zhang, X., Yang, B., Yang, Y., Chen, Q., Wei, Y.: Biocompatible fluorescent organic nanoparticles derived from glucose and polyethylenimine. Colloids Surf., B **123**, 747–752 (2014)
46. Long, Z., Liu, M., Jiang, R., Wan, Q., Mao, L., Wan, Y., Deng, F., Zhang, X., Wei, Y.: Preparation of water soluble and biocompatible AIE-active fluorescent organic nanoparticles via multicomponent reaction and their biological imaging capability. Chem. Eng. J. **308**, 527–534 (2017)
47. Parthasarathy, V., Fery-Forgues, S., Campioli, E., Recher, G., Terenziani, F., Blanchard-Desce, M.: Dipolar versus octupolar triphenylamine-based fluorescent organic nanoparticles as brilliant one-and two-photon emitters for (bio) imaging. Small **7**(22), 3219–3229 (2011)
48. Ishow, E., Brosseau, A., Clavier, G., Nakatani, K., Tauc, P., Fiorini-Debuisschert, C., Neveu, S., Sandre, O., Léaustic, A.: Multicolor emission of small molecule-based amorphous thin films and nanoparticles with a single excitation wavelength. Chem. Mater. **20**(21), 6597–6599 (2008)
49. Amro, K., Daniel, J., Clermont, G., Bsaibess, T., Pucheault, M., Genin, E., Vaultier, M., Blanchard-Desce, M.: A new route towards fluorescent organic nanoparticles with red-shifted emission and increased colloidal stability. Tetrahedron **70**(10), 1903–1909 (2014)
50. Luo, J., Xie, Z., Lam, J.W., Cheng, L., Chen, H., Qiu, C., Kwok, H.S., Zhan, X., Liu, Y., Zhu, D.: Aggregation-induced emission of 1-methyl-1, 2, 3, 4, 5-pentaphenylsilole. Chem. Commun. **18**, 1740–1741 (2001)
51. Zhang, X., Chi, Z., Li, H., Xu, B., Li, X., Zhou, W., Liu, S., Zhang, Y., Xu, J.: Piezofluorochromism of an aggregation-induced emission compound derived from tetraphenylethylene. Chem. Asian J. **6**(3), 808–811 (2011)
52. Zhang, X., Chi, Z., Li, H., Xu, B., Li, X., Liu, S., Zhang, Y., Xu, J.: Synthesis and properties of novel aggregation-induced emission compounds with combined tetraphenylethylene and dicarbazolyl triphenylethylene moieties. J. Mater. Chem. **21**(6), 1788–1796 (2011)
53. Zhang, X., Chi, Z., Zhang, Y., Liu, S., Xu, J.: Recent advances in mechanochromic luminescent metal complexes. J. Mater. Chem. C **1**(21), 3376–3390 (2013)
54. Yu, Y., Feng, C., Hong, Y., Liu, J., Chen, S., Ng, K.M., Luo, K.Q., Tang, B.Z.: Cytophilic fluorescent bioprobes for long-term cell tracking. Adv. Mater. **23**(29), 3298–3302 (2011)
55. Li, Z., Dong, Y.Q., Lam, J.W., Sun, J., Qin, A., Häußler, M., Dong, Y.P., Sung, H.H., Williams, I.D., Kwok, H.S.: Functionalized siloles: versatile synthesis, aggregation-induced emission, and sensory and device applications. Adv. Funct. Mater. **19**(6), 905–917 (2009)
56. An, B.-K., Kwon, S.-K., Jung, S.-D., Park, S.Y.: Enhanced emission and its switching in fluorescent organic nanoparticles. J. Am. Chem. Soc. **124**(48), 14410–14415 (2002)
57. Zhang, X., Yang, Z., Chi, Z., Chen, M., Xu, B., Wang, C., Liu, S., Zhang, Y., Xu, J.: A multi-sensing fluorescent compound derived from cyanoacrylic acid. J. Mater. Chem. **20**(2), 292–298 (2010)
58. Chen, C., Liao, J.-Y., Chi, Z., Xu, B., Zhang, X., Kuang, D.-B., Zhang, Y., Liu, S., Xu, J.: Effect of polyphenyl-substituted ethylene end-capped groups in metal-free organic dyes on performance of dye-sensitized solar cells. RSC Adv. **2**(20), 7788–7797 (2012)

59. Zhang, X., Chi, Z., Zhou, X., Liu, S., Zhang, Y., Xu, J.: Influence of carbazolyl groups on properties of piezofluorochromic aggregation-enhanced emission compounds containing distyrylanthracene. J. Phys. Chem. C **116**(44), 23629–23638 (2012)
60. Zhang, X., Chi, Z., Xu, B., Jiang, L., Zhou, X., Zhang, Y., Liu, S., Xu, J.: Multifunctional organic fluorescent materials derived from 9, 10-distyrylanthracene with alkoxyl endgroups of various lengths. Chem. Commun. **48**(88), 10895–10897 (2012)
61. Bhirde, A., Xie, J., Swierczewska, M., Chen, X.: Nanoparticles for cell labeling. Nanoscale **3**(1), 142–153 (2011)
62. Chi, Z., Zhang, X., Xu, B., Zhou, X., Ma, C., Zhang, Y., Liu, S., Xu, J.: Recent advances in organic mechanofluorochromic materials. Chem. Soc. Rev. **41**(10), 3878–3896 (2012)
63. Zhang, X., Chi, Z., Xu, B., Chen, C., Zhou, X., Zhang, Y., Liu, S., Xu, J.: End-group effects of piezofluorochromic aggregation-induced enhanced emission compounds containing distyrylanthracene. J. Mater. Chem. **22**(35), 18505–18513 (2012)
64. Chen, C., Liao, J.-Y., Chi, Z., Xu, B., Zhang, X., Kuang, D.-B., Zhang, Y., Liu, S., Xu, J.: Metal-free organic dyes derived from triphenylethylene for dye-sensitized solar cells: tuning of the performance by phenothiazine and carbazole. J. Mater. Chem. **22**(18), 8994–9005 (2012)
65. Zhou, X., Li, H., Chi, Z., Zhang, X., Zhang, J., Xu, B., Zhang, Y., Liu, S., Xu, J.: Piezofluorochromism and morphology of a new aggregation-induced emission compound derived from tetraphenylethylene and carbazole. New J. Chem. **36**(3), 685–693 (2012)
66. Zhang, X., Liu, M., Yang, B., Zhang, X., Wei, Y.: Tetraphenylethene-based aggregation-induced emission fluorescent organic nanoparticles: facile preparation and cell imaging application. Colloids Surf., B **112**, 81–86 (2013)
67. Zhang, X., Zhang, X., Yang, B., Zhang, Y., Liu, M., Liu, W., Chen, Y., Wei, Y.: Fabrication of water-dispersible and biocompatible red fluorescent organic nanoparticles via PEGylation of aggregate induced emission enhancement dye and their cell imaging applications. Colloids Surf., B **113**, 435–441 (2014)
68. Zhang, X.-Y., Yang, B.: Facile fabrication of aggregation-induced emission based red fluorescent organic nanoparticles for cell imaging. Chin. J. Polym. Sci. **32**(7), 871–879 (2014)
69. Zhang, X., Zhang, X., Yang, B., Hui, J., Liu, M., Wei, Y.: Facile fabrication of AIE-based stable cross-linked fluorescent organic nanoparticles for cell imaging. Colloids Surf., B **116**, 739–744 (2014)
70. Zhang, X., Zhang, X., Yang, B., Liu, L., Deng, F., Hui, J., Liu, M., Chen, Y., Wei, Y.: Glycosylated aggregation induced emission dye based fluorescent organic nanoparticles: preparation and bioimaging applications. RSC Adv. **4**(46), 24189–24193 (2014)
71. Zhang, X., Zhang, X., Yang, B., Hui, J., Liu, M., Liu, W., Chen, Y., Wei, Y.: PEGylation and cell imaging applications of AIE based fluorescent organic nanoparticles via ring-opening reaction. Polym. Chem. **5**(3), 689–693 (2014)
72. Liu, M., Zhang, X., Yang, B., Liu, L., Deng, F., Zhang, X., Wei, Y.: Polylysine crosslinked AIE dye based fluorescent organic nanoparticles for biological imaging applications. Macromol. Biosci. **14**(9), 1260–1267 (2014)
73. Liu, M., Zhang, X., Yang, B., Deng, F., Yang, Y., Li, Z., Zhang, X., Wei, Y.: Preparation and bioimaging applications of AIE dye cross-linked luminescent polymeric nanoparticles. Macromol. Biosci. **14**(12), 1712–1718 (2014)
74. Xu, D., Zou, H., Liu, M., Tian, J., Huang, H., Wan, Q., Dai, Y., Wen, Y., Zhang, X., Wei, Y.: Synthesis and bioimaging of biodegradable red fluorescent organic nanoparticles with aggregation-induced emission characteristics. J. Colloid Interface Sci. **508**, 248–253 (2017)
75. Zhang, W., Ren, Y.-Y., Zhang, L.-N., Fan, X., Fan, H., Wu, Y., Zhang, Y., Kuang, G.-C.: Boron-difluoride β-diketonate complex as fluorescent organic nanoparticles: aggregation-induced emission for cellular imaging. RSC Adv. **6**(104), 101937–101940 (2016)
76. Zi, L., Meiying, L., Qing, W., Liucheng, M., Hongye, H., Guangjian, Z., Yiqun, W., Fengjie, D., Xiaoyong, Z., Yen, W.: Facile fabrication of PEGylated fluorescent organic nanoparticles with aggregation-induced emission feature via formation of dynamic bonds and their biological imaging applications. Macromol. Rapid Commun. **37**(20), 1657–1661 (2016)

77. Wei, D., Xue, Y., Huang, H., Liu, M., Zeng, G., Wan, Q., Liu, L., Yu, J., Zhang, X., Wei, Y.: Fabrication, self-assembly and biomedical applications of luminescent sodium hyaluronate with aggregation-induced emission feature. Mater. Sci. Eng., C **81**, 120–126 (2017)

78. Xu, D., Liu, M., Zou, H., Huang, Q., Huang, H., Tian, J., Jiang, R., Wen, Y., Zhang, X., Wei, Y.: Fabrication of AIE-active fluorescent organic nanoparticles through one-pot supramolecular polymerization and their biological imaging. J. Taiwan Inst. Chem. Eng. **78**, 455–461 (2017)

79. Xu, D., Liu, M., Zou, H., Tian, J., Huang, H., Wan, Q., Dai, Y., Wen, Y., Zhang, X., Wei, Y.: A new strategy for fabrication of water dispersible and biodegradable fluorescent organic nanoparticles with AIE and ESIPT characteristics and their utilization for bioimaging. Talanta **174**, 803–808 (2017)

80. Jiang, R., Liu, M., Huang, H., Huang, L., Huang, Q., Wen, Y., Cao, Q.-Y., Tian, J., Zhang, X., Wei, Y.: Microwave-assisted multicomponent tandem polymerization for rapid preparation of biodegradable fluorescent organic nanoparticles with aggregation-induced emission feature and their biological imaging applications. Dyes Pigm. **149**, 581–587 (2018)

81. Tang, F., Wang, C., Wang, J., Wang, X., Li, L.: Fluorescent organic nanoparticles with enhanced fluorescence by self-aggregation and their application to cellular imaging. ACS Appl. Mater. Interfaces **6**(20), 18337–18343 (2014)

82. Wang, Z., Yong, T.-Y., Wan, J., Li, Z.-H., Zhao, H., Zhao, Y., Gan, L., Yang, X.-L., Xu, H.-B., Zhang, C.: Temperature-sensitive fluorescent organic nanoparticles with aggregation-induced emission for long-term cellular tracing. ACS Appl. Mater. Interfaces **7**(5), 3420–3425 (2015)

83. Zhang, X., Zhang, X., Yang, B., Zhang, Y., Wei, Y.: A new class of red fluorescent organic nanoparticles: noncovalent fabrication and cell imaging applications. ACS Appl. Mater. Interfaces **6**(5), 3600–3606 (2014)

84. Zhang, X., Zhang, X., Yang, B., Zhang, Y., Wei, Y.: Facile preparation of water dispersible red fluorescent organic nanoparticles and their cell imaging applications. Tetrahedron **70**(22), 3553–3559 (2014)

85. Liu, M., Zhang, X., Yang, B., Deng, F., Ji, J., Yang, Y., Huang, Z., Zhang, X., Wei, Y.: Luminescence tunable fluorescent organic nanoparticles from polyethyleneimine and maltose: facile preparation and bioimaging applications. RSC Adv. **4**(43), 22294–22298 (2014)

86. Liu, M., Zhang, X., Yang, B., Li, Z., Deng, F., Yang, Y., Zhang, X., Wei, Y.: Fluorescent nanoparticles from starch: facile preparation, tunable luminescence and bioimaging. Carbohydr. Polym. **121**, 49–55 (2015)

87. Trofymchuk, K., Reisch, A., Shulov, I., Mély, Y., Klymchenko, A.S.: Tuning the color and photostability of perylene diimides inside polymer nanoparticles: towards biodegradable substitutes of quantum dots. Nanoscale **6**(21), 12934–12942 (2014)

88. Gao, Y., Feng, G., Jiang, T., Goh, C., Ng, L., Liu, B., Li, B., Yang, L., Hua, J., Tian, H.: Biocompatible nanoparticles based on diketo-pyrrolo-pyrrole (DPP) with aggregation-induced red/NIR emission for in vivo two-photon fluorescence imaging. Adv. Funct. Mater. **25**(19), 2857–2866 (2015)

89. Yang, Z., Lee, J.H., Jeon, H.M., Han, J.H., Park, N., He, Y., Lee, H., Hong, K.S., Kang, C., Kim, J.S.: Folate-based near-infrared fluorescent theranostic gemcitabine delivery. J. Am. Chem. Soc. **135**(31), 11657–11662 (2013)

90. Wang, R., Zhang, F.: NIR luminescent nanomaterials for biomedical imaging. J. Mater. Chem. B **2**(17), 2422–2443 (2014)

91. Guo, Z., Park, S., Yoon, J., Shin, I.: Recent progress in the development of near-infrared fluorescent probes for bioimaging applications. Chem. Soc. Rev. **43**(1), 16–29 (2014)

92. Zhang, J., Chen, R., Zhu, Z., Adachi, C., Zhang, X., Lee, C.-S.: Highly stable near-infrared fluorescent organic nanoparticles with a large stokes shift for noninvasive long-term cellular imaging. ACS Appl. Mater. Interfaces **7**(47), 26266–26274 (2015)

93. Iqbal, P.F., Malik, M.A., Wani, W.A.: Serendipity of Cisplatin, and the Emergence of Metallodrugs in Cancer Chemotherapy (2018)

94. Ali, I., Wani, W.A., Saleem, K.: Cancer scenario in India with future perspectives. Cancer Therapy **8** (2011)

95. Ali, I., Wani, W.A., Haque, A., Saleem, K.: Glutamic acid and its derivatives: candidates for rational design of anticancer drugs. Future Med. Chem. **5**(8), 961–978 (2013)

96. Faucon, A., Benhelli-Mokrani, H., Fleury, F., Dutertre, S., Tramier, M., Boucard, J., Lartigue, L., Nedellec, S., Hulin, P., Ishow, E.: Bioconjugated fluorescent organic nanoparticles targeting EGFR-overexpressing cancer cells. Nanoscale **9**(45), 18094–18106 (2017)

97. Xia, Q., Chen, Z., Yu, Z., Wang, L., Qu, J., Liu, R.: Aggregation-induced emission-active near-infrared fluorescent organic nanoparticles for noninvasive long-term monitoring of tumor growth. ACS Appl. Mater. Interfaces **10**(20), 17081–17088 (2018)

98. Prodi, L.: Luminescent chemosensors: from molecules to nanoparticles. New J. Chem. **29**(1), 20–31 (2005)

99. McDonagh, C., Burke, C.S., MacCraith, B.D.: Optical chemical sensors. Chem. Rev. **108**(2), 400–422 (2008)

100. Wang, H., He, F., Yan, R., Wang, X., Zhu, X., Li, L.: Citrate-induced aggregation of conjugated polyelectrolytes for Al^{3+}-ion-sensing assays. ACS Appl. Mater. Interfaces **5**(16), 8254–8259 (2013)

101. Xu, X., Liu, R., Li, L.: Nanoparticles made of π-conjugated compounds targeted for chemical and biological applications. Chem. Commun. **51**(94), 16733–16749 (2015)

102. Ren, C., Zhang, J., Chen, M., Yang, Z.: Self-assembling small molecules for the detection of important analytes. Chem. Soc. Rev. **43**(21), 7257–7266 (2014)

103. Wang, F., Wang, L., Chen, X., Yoon, J.: Recent progress in the development of fluorometric and colorimetric chemosensors for detection of cyanide ions. Chem. Soc. Rev. **43**(13), 4312–4324 (2014)

104. Zhu, C., Liu, L., Yang, Q., Lv, F., Wang, S.: Water-soluble conjugated polymers for imaging, diagnosis, and therapy. Chem. Rev. **112**(8), 4687–4735 (2012)

105. Feng, L., Zhu, C., Yuan, H., Liu, L., Lv, F., Wang, S.: Conjugated polymer nanoparticles: preparation, properties, functionalization and biological applications. Chem. Soc. Rev. **42**(16), 6620–6633 (2013)

106. Wu, M., Xu, X., Wang, J., Li, L.: Fluorescence resonance energy transfer in a binary organic nanoparticle system and its application. ACS Appl. Mater. Interfaces **7**(15), 8243–8250 (2015)

107. Xu, X., Chen, S., Li, L., Yu, G., Liu, Y.: Photophysical properties of polyphenylphenyl compounds in aqueous solutions and application of their nanoparticles for nucleobase sensing. J. Mater. Chem. **18**(22), 2555–2561 (2008)

108. Zhou, J., Yang, Y., Zhang, C.-Y.: Toward biocompatible semiconductor quantum dots: from biosynthesis and bioconjugation to biomedical application. Chem. Rev. **115**(21), 11669–11717 (2015)

109. Venkatesh, V., Shukla, A., Sivakumar, S., Verma, S.: Purine-stabilized green fluorescent gold nanoclusters for cell nuclei imaging applications. ACS Appl. Mater. Interfaces **6**(3), 2185–2191 (2014)

110. Tsang, M.-K., Bai, G., Hao, J.: Stimuli responsive upconversion luminescence nanomaterials and films for various applications. Chem. Soc. Rev. **44**(6), 1585–1607 (2015)

111. Treatment, L.W.: Water Treatment Published by Lenntech. Rotterdamseweg, Netherlands (2004)

112. Hawkes, S.J.: What is a " heavy metal"? J. Chem. Educ. **74**(11), 1374 (1997)

113. Duruibe, J., Ogwuegbu, M., Egwurugwu, J.: Heavy metal pollution and human biotoxic effects. Int. J. Phys. Sci. **2**(5), 112–118 (2007)

114. Crapper, D., Krishnan, S., Dalton, A.: Brain aluminum distribution in Alzheimer's disease and experimental neurofibrillary degeneration. Science **180**(4085), 511–513 (1973)

115. Nayak, P.: Aluminum: impacts and disease. Environ. Res. **89**(2), 101–115 (2002)

116. Walton, J.: An aluminum-based rat model for Alzheimer's disease exhibits oxidative damage, inhibition of PP2A activity, hyperphosphorylated tau, and granulovacuolar degeneration. J. Inorg. Biochem. **101**(9), 1275–1284 (2007)

117. Zhao, Y., Lin, Z., Liao, H., Duan, C., Meng, Q.: A highly selective fluorescent chemosensor for Al^{3+} derived from 8-hydroxyquinoline. Inorg. Chem. Commun. **9**(9), 966–968 (2006)

118. McRae, R., Bagchi, P., Sumalekshmy, S., Fahrni, C.J.: In situ imaging of metals in cells and tissues. Chem. Rev. **109**(10), 4780–4827 (2009)
119. Jiang, Y., Sun, L.L., Ren, G.Z., Niu, X., Wz, Hu, Hu, Z.Q.: A new fluorescence turn-on probe for Aluminum (III) with high selectivity and sensitivity, and its application to bioimaging. ChemistryOpen **4**(3), 378–382 (2015)
120. Huerta-Aguilar, C.A., Raj, P., Thangarasu, P., Singh, N.: Fluorescent organic nanoparticles (FONs) for selective recognition of Al^{3+}: application to bio-imaging for bacterial sample. RSC Adv. **6**(44), 37944–37952 (2016)
121. Kaur, A., Raj, T., Kaur, S., Kaur, N.: Nano molar detection of Al^{3+} in aqueous medium and acidic soil using chromone based fluorescent organic nanoparticles (FONPs). Anal. Methods **6**(21), 8752–8759 (2014)
122. Clifton, J.C.: Mercury exposure and public health. Pediatr. Clin. North America **54**(2), 237, e1-237, e45 (2007)
123. Bjørklund, G.: Mercury and acrodynia. J. Orthomol. Med. **10**(3), 145–146 (1995)
124. Tokuomi, H., Kinoshita, Y., Teramoto, J., Imanishi, K.: Hunter-Russell syndrome. Nihon rinsho. Jpn. J. Clin. Med. **35**, 518–519 (1976)
125. Davidson, P.W., Myers, G.J., Weiss, B.: Mercury exposure and child development outcomes. Pediatrics **113**(Supplement 3), 1023–1029 (2004)
126. James, W.D., Berger, T., Elston, D.: Andrews' Diseases of the Skin: Clinical Dermatology. Elsevier Health Sciences (2015)
127. Sharma, H., Bhardwaj, V.K., Singh, N.: Nanomolar detection of AgI ions in aqueous medium by using naphthalimide-based imine-linked fluorescent organic nanoparticles-application in environmental samples. Eur. J. Inorg. Chem. **2014**(31), 5424–5431 (2014)
128. Delacroix, D., Guerre, J., Leblanc, P., Hickman, C., Penney, B.C.: Radionuclide and radiation protection data handbook. Med. Phys. **30**(2), 277 (2003)
129. Chopra, S., Singh, N., Thangarasu, P., Bhardwaj, V.K., Kaur, N.: Fluorescent organic nanoparticles as chemosensor for nanomolar detection of Cs^+ in aqueous medium. Dyes Pigm. **106**, 45–50 (2014)
130. Ortega, R., Fayard, B., Salomé, M., Devès, G., Susini, J.: Chromium oxidation state imaging in mammalian cells exposed in vitro to soluble or particulate chromate compounds. Chem. Res. Toxicol. **18**(10), 1512–1519 (2005)
131. Eastmond, D.A., MacGregor, J.T., Slesinski, R.S.: Trivalent chromium: assessing the genotoxic risk of an essential trace element and widely used human and animal nutritional supplement. Crit. Rev. Toxicol. **38**(3), 173–190 (2008)
132. Basketter, D., Horev, L., Slodovnik, D., Merimes, S., Trattner, A., Ingber, A.: Investigation of the threshold for allergic reactivity to chromium. Contact Dermatitis **44**(2), 70–74 (2001)
133. Basketter, D., Briatico-Vangosa, G., Kaestner, W., Lally, C., Bontinck, W.: Nickel, cobalt and chromium in consumer products: a role in allergic contact dermatitis? Contact Dermatitis **28**(1), 15–25 (1993)
134. Kaur, N., Kaur, S., Mehan, R., Aguilar, C.A.H., Thangarasu, P., Singh, N.: Fluorescent organic nanoparticles (FONs) of imine-linked peptide for the detection of Cr^{3+} in aqueous medium. Sens. Actuators B: Chem. **206**, 90–97 (2015)
135. Yang, Y., Wang, X., Cui, Q., Cao, Q., Li, L.: Self-assembly of fluorescent organic nanoparticles for Iron (III) sensing and cellular imaging. ACS Appl. Mater. Interfaces **8**(11), 7440–7448 (2016)
136. Azadbakht, R., Hakimi, M., Khanabadi, J.: Fluorescent organic nanoparticles with enhanced fluorescence by self-aggregation and their application for detection of Fe^{3+} ions. New J. Chem. **42**(8), 5929–5936 (2018)
137. Donaldson, J.D., Beyersmann, D.: Cobalt and cobalt compounds. In Ullmann's Encyclopedia of Industrial Chemistry, (Ed.) (2005). doi:https://doi.org/10.1002/14356007.a07_281.pub2
138. Morin, Y., Tĕtu, A., Mercier, G.: Québec beer-drinkers' cardiomyopathy: clinical and hemodynamic aspects. Ann. N. Y. Acad. Sci. **156**(1), 566–576 (1969)
139. Barceloux, D.G., Barceloux, D.: Cobalt. J. Toxicol. Clin. Toxicol. **37**(2), 201–216 (1999)

140. Basketter, D.A., Angelini, G., Ingber, A., Kern, P.S., Menné, T.: Nickel, chromium and cobalt in consumer products: revisiting safe levels in the new millennium. Contact Dermatitis **49**(1), 1–7 (2003)

141. Mahajan, P.G., Dige, N.C., Desai, N.K., Patil, S.R., Kondalkar, V.V., Hong, S.-K., Lee, K.H.: Selective detection of Co^{2+} by fluorescent nano probe: diagnostic approach for analysis of environmental samples and biological activities. Spectrochim. Acta Part A Mol. Biomol. Spectrosc. **198**, 136–144 (2018)

142. Kaur, G., Singh, A., Venugopalan, P., Kaur, N., Singh, N.: Selective recognition of lithium (i) ions using Biginelli based fluorescent organic nanoparticles in an aqueous medium. RSC Adv. **6**(3), 1792–1799 (2016)

143. Fosmire, G.J.: Zinc toxicity. Am. J. Clin. Nutr. **51**(2), 225–227 (1990)

144. Mallevialle, J., Bruchet, A., Fiessinger, F.: How safe are organic polymers in water treatment? J. Am. Water Works Assoc. 87–93 (1984)

145. Li, S., Zhou, Y., Yu, C., Chen, F., Xu, J.: Switching the ligand-exchange reactivities of chloro-bridged cyclopalladated azobenzenes for the colorimetric sensing of thiocyanate. New J. Chem. **33**(7), 1462–1465 (2009)

146. Bhardwaj, S., Maurya, N., Singh, A.K.: Chromone based fluorescent organic nanoparticles for high-precision in-situ sensing of Cu^{2+} and CN^- ions in 100% aqueous solutions. Sens. Actuators B: Chem. **260**, 753–762 (2018)

147. Baldessarini, R.J., Tondo, L., Davis, P., Pompili, M., Goodwin, F.K., Hennen, J.: Decreased risk of suicides and attempts during long-term lithium treatment: a meta-analytic review. Bipolar Disord. **8**(5), 625–639 (2006)

148. Pettilä, V., Takkunen, O., Tukiainen, P.: Zinc chloride smoke inhalation: a rare cause of severe acute respiratory distress syndrome. Intensive Care Med. **26**(2), 215–217 (2000)

149. Huerta-Aguilar, C.A., Pandiyan, T., Raj, P., Singh, N., Zanella, R.: Fluorescent organic nanoparticles (FONs) for the selective recognition of Zn^{2+}: applications to multi-vitamin formulations in aqueous medium. Sens. Actuators B: Chem. **223**, 59–67 (2016)

150. Letterman, R.D., Pero, R.W.: Contaminants in polyelectrolytes used in water treatment. J. Am. Water Works Assoc. 87–97 (1990)

151. Lakshminarayanan, P., Suresh, E., Ghosh, P.: Synthesis and characterization of a tripodal amide ligand and its binding with anions of different dimensionality. Inorg. Chem. **45**(11), 4372–4380 (2006)

152. Xue, W., Li, L., Li, Q., Wu, A.: Novel furo [2, 3-d] pyrimidine derivative as fluorescent chemosensor for HSO^{4-}. Talanta **88**, 734–738 (2012)

153. Spence, G.T., Chan, C., Szemes, F., Beer, P.D.: Anion binding induced conformational changes exploited for recognition, sensing and pseudorotaxane disassembly. Dalton Trans. **41**(43), 13474–13485 (2012)

154. Kato, R., Kawai, A., Hattori, T.: Optical detection of anions using N-(4-(4-nitrophenylazo) phenyl)-N′-propyl thiourea bound silica film. New J. Chem. **37**(3), 717–721 (2013)

155. Luo, Y.-H., Ge, S.-W., Song, W.-T., Sun, B.-W.: Supramolecular assembly and host–guest interaction of crown ether with inorganic acid and organic amine containing carboxyl groups. New J. Chem. **38**(2), 723–729 (2014)

156. Santos-Figueroa, L.E., Moragues, M.E., Climent, E., Agostini, A., Martínez-Máñez, R., Sancenón, F.: Chromogenic and fluorogenic chemosensors and reagents for anions. A comprehensive review of the years 2010–2011. Chem. Soc. Rev. **42**(8), 3489–3613 (2013)

157. Lee, G.W., Singh, N., Jung, H.J., Jang, D.O.: Selective anion recognition by retarding the cooperative polarization effect of amide linkages. Tetrahedron Lett. **50**(7), 807–810 (2009)

158. Kumar, M.S., Kumar, S.L.A., Sreekanth, A.: Highly selective fluorogenic anion chemosensors: naked-eye detection of F^- and AcO^- ions in natural water using a test strip. Anal. Methods **5**(22), 6401–6410 (2013)

159. Lou, X., Ou, D., Li, Q., Li, Z.: An indirect approach for anion detection: the displacement strategy and its application. Chem. Commun. **48**(68), 8462–8477 (2012)

160. Kim, H.J., Bhuniya, S., Mahajan, R.K., Puri, R., Liu, H., Ko, K.C., Lee, J.Y., Kim, J.S.: Fluorescence turn-on sensors for HSO_4^-. Chem. Commun. **46**, 7128–7130 (2009)

161. Lu, W., Zhou, J., Liu, K., Chen, D., Jiang, L., Shen, Z.: A polymeric film probe with a turn-on fluorescence response to hydrogen sulfate ions in aqueous media. J. Mater. Chem. B **1**(38), 5014–5020 (2013)

162. Chopra, S., Singh, J., Singh, N., Kaur, N.: Fluorescent organic nanoparticles of tripodal receptor as sensors for HSO_4^- in aqueous medium: application to real sample analysis. Anal. Methods **6**(22), 9030–9036 (2014)

163. Ward, J.M., Ohshima, M.: The role of iodine in carcinogenesis. In: Essential Nutrients in Carcinogenesis, pp. 529–542. Springer, 1986

164. Kaur, A., Raj, T., Kaur, S., Singh, N., Kaur, N.: Fluorescent organic nanoparticles of dihydropyrimidone derivatives for selective recognition of iodide using a displacement assay: application of the sensors in water and biological fluids. Org. Biomol. Chem. **13**(4), 1204–1212 (2015)

165. Chopra, S., Singh, J., Kaur, H., Singh, H., Singh, N., Kaur, N.: Selective chemosensing of spermidine based on fluorescent organic nanoparticles in aqueous media via a Fe^{3+} displacement assay. New J. Chem. **39**(5), 3507–3512 (2015)

166. Chopra, S., Singh, J., Kaur, H., Singh, N., Kaur, N.: Estimation of biogenic amines and biothiols by metal complex of fluorescent organic nanoparticles acting as single receptor multi-analyte sensor in aqueous medium. Sens. Actuators B: Chem. **220**, 295–301 (2015)

167. Kaur, N., Kaur, M., Chopra, S., Singh, J., Kuwar, A., Singh, N.: Fe(III) conjugated fluorescent organic nanoparticles for ratiometric detection of tyramine in aqueous medium: a novel method to determine food quality. Food Chem. **245**, 1257–1261 (2018)

168. Bhardwaj, V.K., Sharma, H., Singh, N.: Ratiometric fluorescent probe for biothiol in aqueous medium with fluorescent organic nanoparticles. Talanta **129**, 198–202 (2014)

169. Hu, S., Huang, Q., Lin, Y., Wei, C., Zhang, H., Zhang, W., Guo, Z., Bao, X., Shi, J., Hao, A.: Reduced graphene oxide-carbon dots composite as an enhanced material for electrochemical determination of dopamine. Electrochim. Acta **130**, 805–809 (2014)

170. Venton, B.J., Wightman, R.M.: Psychoanalytical electrochemistry: dopamine and behavior. Anal. Chem. **75**(19), 414 A–421 A (2003)

171. Iqbal, Z., Lai, E.P.C., Avis, T.J.: Antimicrobial effect of polydopamine coating on *Escherichia coli*. J. Mater. Chem. **22**(40), 21608–21612 (2012)

172. Ding, L., Qin, Z., Xiang, C., Zhou, G.: Novel fluorescent organic nanoparticles as a label-free biosensor for dopamine in serum. J. Mater. Chem. B **5**(15), 2750–2756 (2017)

173. Bieri, M., Bürgi, T.: d-Penicillamine adsorption on gold: an in situ ATR-IR spectroscopic and QCM study. Langmuir **22**(20), 8379–8386 (2006)

174. Kean, W., Howard-Lock, H., Lock, C.: Chirality in antirheumatic drugs. The Lancet **338**(8782–8783), 1565–1568 (1991)

175. Saracino, M.A., Cannistraci, C., Bugamelli, F., Morganti, E., Neri, I., Balestri, R., Patrizi, A., Raggi, M.A.: A novel HPLC-electrochemical detection approach for the determination of d-penicillamine in skin specimens. Talanta **103**, 355–360 (2013)

176. Mahajan, P.G., Kolekar, G.B., Patil, S.R.: Recognition of D-Penicillamine using Schiff base centered fluorescent organic nanoparticles and application to medicine analysis. J. Fluoresc. **27**(3), 829–839 (2017)

177. Food, Drug Administration, H., Food Labeling: Revision of the nutrition and supplement facts labels. Final rule. Fed. Reg. **81**(103), 33741–33999 (2016)

178. Mahajan, P.G., Dige, N.C., Suryawanshi, S.B., Dalavi, D.K., Kamble, A.A., Bhopate, D.P., Kadam, A.N., Kondalkar, V.V., Kolekar, G.B., Patil, S.R.: FRET Between Riboflavin and 9-Anthraldehyde based fluorescent organic nanoparticles possessing antibacterial activity. J. Fluoresc. **28**(1), 207–215 (2018)

179. Diaz, M.H., Hauser, A.R.: *Pseudomonas aeruginosa* cytotoxin ExoU is injected into phagocytic cells during acute pneumonia. Infect. Immun. **78**(4), 1447–1456 (2010)

180. Fazeli, H., Akbari, R., Moghim, S., Narimani, T., Arabestani, M.R., Ghoddousi, A.R.: *Pseudomonas aeruginosa* infections in patients, hospital means, and personnel's specimens. J. Res. Med. Sci. **17**(4) (2012)

181. Seki, M., Machida, N., Yamagishi, Y., Yoshida, H., Tomono, K.: Nosocomial outbreak of multidrug-resistant *Pseudomonas aeruginosa* caused by damaged transesophageal echocardiogram probe used in cardiovascular surgical operations. J. Infect. Chemother. **19**(4), 677–681 (2013)

182. Tam, V.H., Chang, K.-T., Abdelraouf, K., Brioso, C.G., Ameka, M., McCaskey, L.A., Weston, J.S., Caeiro, J.-P., Garey, K.W.: Prevalence, resistance mechanisms, and susceptibility of multidrug-resistant bloodstream isolates of *Pseudomonas aeruginosa*. Antimicrob. Agents Chemother. **54**(3), 1160–1164 (2010)

183. Lee, T.W., Brownlee, K.G., Denton, M., Littlewood, J.M., Conway, S.P.: Reduction in prevalence of chronic *Pseudomonas aeruginosa* infection at a regional pediatric cystic fibrosis center. Pediatr. Pulmonol. **37**(2), 104–110 (2004)

184. Lyczak, J.B., Cannon, C.L., Pier, G.B.: Lung infections associated with cystic fibrosis. Clin. Microbiol. Rev. **15**(2), 194–222 (2002)

185. Lyczak, J.B., Cannon, C.L., Pier, G.B.: Establishment of *Pseudomonas aeruginosa* infection: lessons from a versatile opportunist. Microbes Infect. **2**(9), 1051–1060 (2000)

186. Lister, P.D., Wolter, D.J., Hanson, N.D.: Antibacterial-resistant *Pseudomonas aeruginosa*: clinical impact and complex regulation of chromosomally encoded resistance mechanisms. Clin. Microbiol. Rev. **22**(4), 582–610 (2009)

187. Landman, D., Bratu, S., Kochar, S., Panwar, M., Trehan, M., Doymaz, M., Quale, J.: Evolution of antimicrobial resistance among *Pseudomonas aeruginosa*, *Acinetobacter baumannii* and *Klebsiella pneumoniae* in Brooklyn. N. Y. J. Antimicrob. Chemother. **60**(1), 78–82 (2007)

188. Kaur, G., Raj, T., Kaur, N., Singh, N.: Pyrimidine-based functional fluorescent organic nanoparticle probe for detection of *Pseudomonas aeruginosa*. Org. Biomol. Chem. **13**(16), 4673–4679 (2015)

189. Tiwari, G., Tiwari, R., Sriwastawa, B., Bhati, L., Pandey, S., Pandey, P., Bannerjee, S.K.: Drug delivery systems: an updated review. Int. J. Pharm. Invest. **2**(1), 2 (2012)

190. Jana, A., Nguyen, K.T., Li, X., Zhu, P., Tan, N.S., Ågren, H., Zhao, Y.: Perylene-derived single-component organic nanoparticles with tunable emission: efficient anticancer drug carriers with real-time monitoring of drug release. ACS Nano **8**(6), 5939–5952 (2014)

191. Breton, M., Prével, G., Audibert, J.-F., Pansu, R., Tauc, P., Le Pioufle, B., Français, O., Fresnais, J., Berret, J.-F., Ishow, E.: Solvatochromic dissociation of non-covalent fluorescent organic nanoparticles upon cell internalization. Phys. Chem. Chem. Phys. **13**(29), 13268–13276 (2011)

192. Yavuz, M.S., Cheng, Y., Chen, J., Cobley, C.M., Zhang, Q., Rycenga, M., Xie, J., Kim, C., Song, K.H., Schwartz, A.G.: Gold nanocages covered by smart polymers for controlled release with near-infrared light. Nat. Mater. **8**(12), 935–939 (2009)

193. Jana, A., Devi, K.S.P., Maiti, T.K., Singh, N.P.: Perylene-3-ylmethanol: fluorescent organic nanoparticles as a single-component photoresponsive nanocarrier with real-time monitoring of anticancer drug release. J. Am. Chem. Soc. **134**(18), 7656–7659 (2012)

194. Jana, A., Saha, B., Banerjee, D.R., Ghosh, S.K., Nguyen, K.T., Ma, X., Qu, Q., Zhao, Y., Singh, N.P.: Photocontrolled nuclear-targeted drug delivery by single component photoresponsive fluorescent organic nanoparticles of acridin-9-methanol. Bioconjug. Chem. **24**(11), 1828–1839 (2013)

195. Gangopadhyay, M., Singh, T., Behara, K.K., Karwa, S., Ghosh, S., Singh, N.P.: Coumarin-containing-star-shaped 4-arm-polyethylene glycol: targeted fluorescent organic nanoparticles for dual treatment of photodynamic therapy and chemotherapy. Photochem. Photobiol. Sci. **14**(7), 1329–1336 (2015)

196. Barman, S., Mukhopadhyay, S.K., Behara, K.K., Dey, S., Singh, N.P.: 1-acetylpyrene–salicylic acid: photoresponsive fluorescent organic nanoparticles for the regulated release of a natural antimicrobial compound, salicylic acid. ACS Appl. Mater. Interfaces **6**(10), 7045–7054 (2014)

197. Bonnett, R.: Chemical aspects of photodynamic therapy. CRC Press, Boca Raton (2000)

198. Detty, M.R., Gibson, S.L., Wagner, S.J.: Current clinical and preclinical photosensitizers for use in photodynamic therapy. J. Med. Chem. **47**(16), 3897–3915 (2004)

199. Wani, W.A., Baig, U., Shreaz, S., Shiekh, R.A., Iqbal, P.F., Jameel, E., Ahmad, A., Mohd-Setapar, S.H., Mushtaque, M., Hun, L.T.: Recent advances in iron complexes as potential anticancer agents. New J. Chem. (2016)

200. Chang, C.-C., Hsieh, M.-C., Lin, J.-C., Chang, T.-C.: Selective photodynamic therapy based on aggregation-induced emission enhancement of fluorescent organic nanoparticles. Biomaterials 33(3), 897–906 (2012)

201. Austin, L.A., Kang, B., Yen, C.-W., El-Sayed, M.A.: Plasmonic imaging of human oral cancer cell communities during programmed cell death by nuclear-targeting silver nanoparticles. J. Am. Chem. Soc. 133(44), 17594–17597 (2011)

202. Srinivas, P., Patra, C.R., Bhattacharya, S., Mukhopadhyay, D.: Cytotoxicity of naphthoquinones and their capacity to generate reactive oxygen species is quenched when conjugated with gold nanoparticles. Int J Nanomed. 6, 2113–2122 (2011)

203. Sage, J., Van Uitert, R., Duffy, T.: Early changes in blood brain barrier permeability to small molecules after transient cerebral ischemia. Stroke 15(1), 46–50 (1984)

204. Kuroiwa, T., Ting, P., Martinez, H., Klatzo, I.: The biphasic opening of the blood-brain barrier to proteins following temporary middle cerebral artery occlusion. Acta Neuropathol. 68(2), 122–129 (1985)

205. Brightman, M.W., Klatzo, I., Olsson, Y., Reese, T.S.: The blood-brain barrier to proteins under normal and pathological conditions. J. Neurol. Sci. 10(3), 215–239 (1970)

206. Stoll, G., Kleinschnitz, C., Meuth, S.G., Braeuninger, S., Ip, C.W., Wessig, C., Nölte, I., Bendszus, M.: Transient widespread blood—brain barrier alterations after cerebral photothrombosis as revealed by gadofluorine M-enhanced magnetic resonance imaging. J. Cereb. Blood Flow Metab. 29(2), 331–341 (2009)

207. Latour Lawrence, L., Kang, D.-W., Ezzeddine Mustapha, A., Chalela Julio, A., Warach, S.: Early blood–brain barrier disruption in human focal brain ischemia. Ann. Neurol. 56(4), 468–477 (2004)

208. Khatri, R., McKinney, A.M., Swenson, B., Janardhan, V.: Blood–brain barrier, reperfusion injury, and hemorrhagic transformation in acute ischemic stroke. Neurology 79(13 Supplement 1), S52–S57 (2012)

209. Cai, X., Bandla, A., Mao, D., Feng, G., Qin, W., Liao, L.-D., Thakor, N., Tang Ben, Z., Liu, B.: Biocompatible red fluorescent organic nanoparticles with tunable size and aggregation-induced emission for evaluation of blood-brain barrier damage. Adv. Mater. 28(39), 8760–8765 (2016)

210. Shilo, M., Sharon, A., Baranes, K., Motiei, M., Lellouche, J.-P.M., Popovtzer, R.: The effect of nanoparticle size on the probability to cross the blood-brain barrier: an in-vitro endothelial cell model. J. Nanobiotechnol. 13(1), 19 (2015)

211. Labat-gest, V., Tomasi, S.: Photothrombotic ischemia: a minimally invasive and reproducible photochemical cortical lesion model for mouse stroke studies. J. Visualized Exp.: JoVE 76, 50370 (2013)

212. Talley Watts, L., Zheng, W., Garling, R.J., Frohlich, V.C., Lechleiter, J.D.: Rose Bengal photothrombosis by confocal optical imaging in vivo: a model of single vessel stroke. J. Visualized Exp.: JoVE 100, 52794 (2015)

Chapter 4
Conclusions and Future Outlook

The main focus of this book has been to highlight the recent progress in the applications of FONs as cell imaging and chemosensing agents and drug delivery systems. Basically, four important methods including self-assembly, polymerization, emulsification and nanoprecipitation/reprecipitation have been used for the preparation of FONs with diverse applications in analytical and biomedical sciences. Out of these techniques, nanoprecipitation is the simplest and the most widely used technique. This technique enables the transformation of soluble organic molecules into nanoparticles in the aqueous media and later ensures their fast screening for various analytical and biomedical applications. In the context of fluorescence-based sensing, drug delivery and cell imaging, the properties of FONs that are of major importance are stability, brightness, toxicity and biodegradability.

The recent research carried out on the design and development of AIE nanoparticles is a promising stimulation towards the development of highly fluorescent nanoparticles as the natural aggregation is kept busy in increasing the fluorescence of the nanoparticles. AIE is, therefore, truly a novel finding of this subject. The encapsulation of an emitter into different matrices affects their aggregation, molecular packing and distribution in the nanoparticles. Thus, care is needed while selecting the polymeric matrix and the emitter to matrix ratio (Fig. 4.1), which may help in the tuning of nanoparticle size, brightness and stability. Several reports have been published with polymer encapsulated emitters having sizes ranging from few to several hundred nanometers. Such nanoparticles have versatile surface functional groups that have been tailored to provide space for different imaging needs, for example, imaging of cellular organelles, targeted in vitro and in vivo imaging of tumours, tracing of cancer cells, and imaging of blood vessels and specific chemical and biomolecular species. The photostability of FONs is mainly dependent on the type of the emitter used. However, aggregation into nanoparticles regime improves their stability in the context of photobleaching. The more important advantages of FONs are their excellent biocompatibilities, enhanced cellular permeabilities and low cytotoxicities in vitro. However, their systematic in vivo toxicity, biodistribution and biodegradability is yet to be explored to arrive at solid conclusions (Fig. 4.1).

© The Author(s), under exclusive license to Springer Nature Singapore Pte Ltd. 2018
W. A. Wani et al., *Fluorescent Organic Nanoparticles*, SpringerBriefs in Materials,
https://doi.org/10.1007/978-981-13-2655-4_4

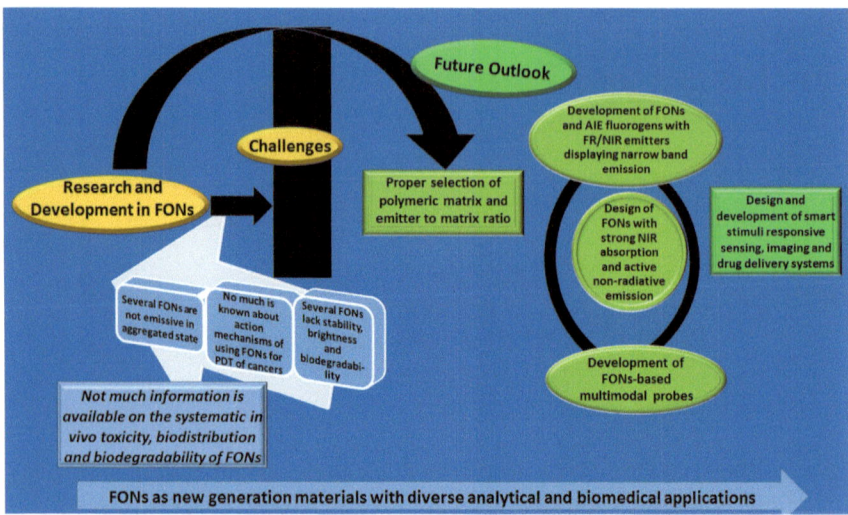

Fig. 4.1 An overview of the challenges and future outlook of research in the development of FONs for analytical and biomedical applications

The recent exploration of the cell imaging, chemosensing and drug delivery applications of FONs has brought about a great revolution. This is very important and interesting from the materials point of view. Irrespective of the fact that there has been a good deal of work for the development of highly fluorescent nanoparticles for cell imaging, sensing and drug delivery applications, several existing FONs are not very emissive in the aggregated states. Thus, FONs with strong NIR absorption and active non-radiative emission are the materials of choice at the moment (Fig. 4.1). The non-radiative pathway is generally associated with heat production, and thus, such FONs might have great potential for use in photothermal therapy (PTT), wherein highly specific, non-toxic and non-invasive treatments of cancers may be carried out [1]. It is of great interest that several FONs have also been tried as agents for photodynamic therapy of cancers; however, this field of research is at an early stage because the principles of design of such FONs and their action mechanisms are not fully established (Fig. 4.1). Nevertheless, such nanoparticulate systems have successfully confirmed their promise for the deep in vivo tumour imaging and PDT with minimum or no side effects on normal cells and tissues.

The future research on FONs needs to be focused on the design and development of smart stimuli-responsive sensing, imaging and drug delivery systems (Fig. 4.1). Besides, the future work must focus on developing FONs and AIE fluorogens with FR/NIR emitters displaying narrow band emission for specific analytical and biomedical applications (Fig. 4.1). Several strategies for signal amplification strategies including FRET and metal-enhanced fluorescence can further improve brightness for fluorescence imaging. On similar terms, the design and development of FONs based multimodal probes will be perfect for imaging applications both in vitro and in vivo. This may make accessible higher sensitivity and accuracy of the complementary information provided by each imaging modality. A focused and proper design must give birth to single FONs with concerted ability for fluorescence imaging and PDT effect. This must provide extraordinary benefits in creating single component multimodality image-guided therapeutic agents. For the fine-tuning of the size and fluorescence of such FONs, the structural features including molecular functionalities and topology must be properly developed that might help in the delivery of the imaging and therapeutic agents. As an example, FONs with the ability to sense an analyte, bind/load or deliver a drug, and undergo fluorescence changes specific to a cellular environment in the presence of light, enzyme or pH must be developed for targeted and responsive sensing, drug delivery and cancer cell imaging applications, respectively. Overall, FONs represent a very interesting field of research with promise for varied applications in analytical and biomedical sciences.

Reference

1. Xu, L., Cheng, L., Wang, C., Peng, R., Liu, Z.: Conjugated polymers for photothermal therapy of cancer. Polym. Chem. **5**(5), 1573–1580 (2014)